THE HIPPIES WERE RIGHT!

ABOUT CONSCIOUSNESS AND EVERYTHING ELSE

Tamlyn Merryan Hunt

imprint-academic.com

Published in the UK by
Imprint Academic, PO Box 200, Exeter EX5 5YX, UK

Distributed in the USA by
Lightning Source,
La Vergne, TN 37086, USA

ISBN 9781788361415 Paperback

A CIP catalogue record for this book is available from the
British Library and US Library of Congress

This product complies with the General Product Safety Regulations
introduced by the European Union in 2024.

EU GPSR Authorised Representative:
LOGOS EUROPE, 9 rue Nicolas Poussin,
17000, LA ROCHELLE, France

E-mail: Contact@logoseurope.eu

Contents

Preface

Vibrations, Man...

In the beginning was the word, and the word was *Ommmmmm.* Vibrating in the throat and spreading down into the body and up into the head.

That's something my father might have said in his extended moments of caffeine- and nicotine-fueled eloquence. He claimed to have helped birth the hippie movement both in England, where he was born (as was I), and later in the U.S., bringing with him a potent blend of rebellious politics, Eastern mysticism, psychedelic exploration, and countercultural values that defined an era.

Growing up as the son of a charismatic iconoclast, a traveling philosopher, and occasional drug dealer, my father, along with his handicraft-loving back-to-the-land black sheep wife, my mother, planted seeds in my consciousness that would take decades to fully sprout—seeds that would eventually grow into the philosophical and scientific framework that is the main topic of this book. I'm a black sheep son of two black sheep (which makes me a rainbow sheep?).

As a child of the counterculture, I naturally rebelled against my hippie roots in my teen years. The pendulum swings both ways. I joined the U.S. Army a year after high school and didn't have much patience for spirituality or religion. I was a militant atheist for a while, embracing science and materialism and the death of God and all that. While many of my peers were growing their hair long and questioning authority, I cut mine short and, once my Army years were over, dove into the structured world of science and academia.

Photo: My hippie family, with VW bus, in 1975 England, rocking the bell bottoms.

Yet even as I distanced myself from what seemed to me like pseudoscientific babble about cosmic vibrations and universal consciousness, I couldn't shake the feeling that beneath the tie-dye and patchouli maybe there was something genuine—an intuition about reality that warranted serious investigation rather than dismissal.

My twenties were a period of reconciliation. I had been deeply philosophical starting in my teens, and even though I had rejected much of the counterculture at that time, I read very widely during my Army years, particularly in philosophy, science, and the classics. Even though it was a job where I would have been thrown in jail if I had tried to quit (yes, that's the consequence of going "AWOL"), what I lacked in freedom I gained in free time. And I used that free time to read. A lot. I plowed through a couple of books a week for four years, establishing a great foundation for my later more directed studies in biology, environmental science, and law.

I found myself drawn back to those counterculture roots after leaving the military, at 23 years old, smoking cannabis and experimenting with psychedelics like LSD, psilocybin mushrooms, and eventually more powerful psychedelics and entheogens. These substances allowed me to experience directly what had previously been abstract concepts. Watching patterns emerge and dissolve, feeling the boundaries between self and world temporarily disappear, I began to wonder: what if these experiences weren't merely chemical hallucinations but glimpses of a deeper reality that science liked to ignore? I was, of course, far from alone in such questions.

These questions set me on a path that would eventually lead to collaboration with researchers like Professor Jonathan Schooler at UC Santa Barbara's METALAB, my "academic daddy" as I like to call him. Together, we've been developing over the last dozen or so years what we call the General Resonance Theory (GRT) of consciousness—which proposes that consciousness arises from synchronized oscillations (resonance) in physical systems. Far from the vague pronouncements of hippie mysticism, GRT is a scientifically rigorous attempt to address one of the most challenging problems in science: how physical processes give rise to subjective experience. Indeed, the modern term for this problem is the "hard problem" of consciousness, distinguished from all of those "easy problems" of neuroscience and its focus on the workings of neurons and the brain. (Yes, "easy" is meant to be entirely tongue in cheek.)

The central insight of GRT is quite simple: when things resonate in proximity, they create something new and greater than the sum of their parts. Think about music—individual notes played together create chords and harmonies that have qualities none of the individual notes possess, based on the relationships between the notes, which is something above and beyond the parts. In a similar way, we think the countless tiny vibrations in your brain—electromagnetic fields organized in nested and synchronized patterns—are the symphony of your conscious experience. Yes, *are*, not "correlate with" or "cause" consciousness—they *are* my, and your, consciousness.

This insight isn't just poetic; it's showing up in cutting-edge neuroscience. When you're conscious, your brain displays specific patterns of synchronized electrical activity. When these patterns are disrupted—during deep sleep, under anesthesia, or during certain types of seizures—consciousness changes dramatically or disappears altogether. The relationship between synchronized vibrations and consciousness seems too consistent to be coincidental.

What makes our approach unique is that it offers a potential solution to a problem philosophers call the "combination problem" —how simpler forms of consciousness might combine to form more complex ones. By showing how vibrating elements can synchronize in complex nested patterns while still maintaining their individuality, GRT provides a model for understanding how consciousness might work at different levels, from the simplest organisms to complex human awareness.

This model doesn't just apply to brains. Throughout nature, we see resonance—synchronized vibration—creating organization and complexity. From the molecular vibrations that allow photosynthesis to capture sunlight, to the synchronous flashing of fireflies, to the coordinated movements of bird flocks and fish schools, resonance appears to be a fundamental organizing principle in our universe. Not all resonating structures are conscious, but all consciousness is based on various kinds of resonance. For example, the resonating group of fireflies or bird flocks don't have their own higher level of consciousness in the group, but the resonating systems that comprise the individual firefly or the bird itself do enjoy a relatively complex consciousness.

And no, rocks are not conscious—I'll delve into the details of what's known as "panpsychism" later in this book. This is the idea that all matter has some associated consciousness, albeit a very rudimentary kind of consciousness in most forms of matter. However, even if we suggest, as our approach does, that indeed there is some tiny iota of consciousness even in simple forms of matter, it's important to understand at what *level* of organization that consciousness resides. The rock is not conscious, but the atoms that comprise it may have that tiny something at their level of organization.

This view on the nature of reality and consciousness brings us back to the insights of ancient wisdom traditions, many of which described reality in terms of vibration and resonance, long before modern science. The Hindu sacred sound of OM (say this and feel your throat and voicebox vibrate actively as you do) as the primordial sound from which reality manifests, the Buddhist understanding of impermanence and interdependence, and even the biblical "In the beginning was the Word," all hint at a vibrational understanding of reality that modern physics is now moving towards in various ways.

The implications extend far beyond neuroscience. In physics, we've discovered that reality at its most fundamental level consists not of solid particles but of vibrating fields. In biology, we're finding that resonance enables the self-organization of complex systems across multiple scales. In technology, as we create increasingly sophisticated AI systems that mimic aspects of our own intelligence, we face profound questions about whether these systems could ever resonate in ways that support consciousness. Are entirely different computer architectures required for AI to be in any way conscious? I argue that the answer is probably "yes." Maybe?

Each chapter of this book explores a different aspect of this resonance-based approach to consciousness. We'll examine the historical "hard problem" of consciousness and how resonance theory offers a new perspective based on panpsychism, the notion that all matter enjoys at least a tiny bit of consciousness. We'll dive into the physics of waves and fields, exploring how resonance operates at different scales of organization. We'll look at how electromagnetic fields in the brain might be the actual medium of conscious experience, not just a byproduct of neural activity. We'll consider how these ideas might help us understand the full spectrum of conscious experience, from everyday awareness to meditation and psychedelic states.

I weave in my own journey, psychedelics and all, as well as many of the books I read over the decades that had important impacts on my thinking. I also focus on "flower power" and the resonant frequencies of peace! Are we seeing the dawning of a new age of peace, driven by integrating the lessons of peace and love that surfaced strongly in the 60s and 70s and helped to end the Vietnam War? Or are we irrevocably on the path of war as American empire extends further and further into the world, with all of the militarized violence that has entailed over recent decades?

We'll also tackle perhaps the most timely question of all: could artificial intelligence systems ever become conscious? What kind of future can we expect as AIs become like gods? As AI models like ChatGPT, Claude, and Gemini demonstrate increasingly sophisticated capabilities, understanding what makes something conscious becomes not just a philosophical puzzle but a pressing ethical and safety concern. If consciousness is fundamentally about resonant integration, what would that mean for how we design, use, and relate to artificial systems?

My journey from hippie kid to atheist skeptic to psychonaut to philosopher, peace activist, and scientific researcher has come full circle. The same intuitions that my parents and their generation expressed often in mystical terms—that consciousness is fundamental, that "everything vibrates," that reality is interconnected—are finding validation in cutting-edge scientific research. The hippies, it turns out, were onto something profound. It might even be fair to say they were right about a lot of things. Maybe not *everything* (despite my title), but yeah a lot of things.

This doesn't mean we should uncritically embrace every aspect of counterculture philosophy. Many specific claims made during that era remain unsupported by evidence. But the core intuition—that vibration, resonance, and coherence are central to understanding consciousness and its place in the physical world—deserves serious scientific attention. Indeed, it is receiving significant scientific attention around the world now. But these ideas aren't important "only" to science and philosophy. They're also profoundly practical.

While my father preached cosmic vibrations and universal consciousness with tie-dyed flair, my mother embodied a different, equally profound dimension of the counterculture. Her quiet perseverance, spinning wool into yarn with practiced hands, knitting sweaters and hats that kept us warm through winter, and tending our home when my father's "spiritual quests" took him elsewhere, taught me lessons about resonance that were more grounded but no less profound. If my father represented the theoretical physics of hippie philosophy, my mother was its practical engineering—transforming abstract ideals into the tangible necessities of daily life. Her hands, constantly in steady and deliberate motion, creating order and beauty from raw materials, demonstrated a kind of embodied resonance that would later inform my understanding of how consciousness operates in physical systems.

Photo: My mother and the three "big kids" in our family in 1975 (my youngest sister came much later, with a different father). I'm on the left sucking my finger.

She told me recently about a friend in England who marveled at her many pursuits. "A friend in England years ago said she'd never met anyone with so many hobbies, but I thought of them as life skills, not hobbies." This distinction—between what mainstream society saw as eccentric lifestyle choices and what practitioners understood as essential knowledge—captures one of the major misunderstandings about the hippie movement. What appeared to outsiders as recreational drug use, impractical communalism, and mystical nonsense was often a serious exploration of consciousness, sustainability, and alternative ways of organizing human life—coming from a recognition that modern society simply doesn't work very well for people or planet, in so many ways. The hippies weren't just dropping out; they were developing life skills for a different kind of civilization. And as our current civilization faces mounting crises—ecological, technological, moral, and spiritual—their insights about vibration, consciousness, lifestyles, and love may prove to be exactly the life skills we need for our next era.

There was of course a darker side to the hippie movement that we can't ignore. Many of its leaders were profoundly hypocritical and irresponsible. Some were predatory. I discuss this shadow side too.

My own brief childhood experience in the Love Family—one of the larger counterculture communes of the era, based in Washington State—taught me firsthand that beautiful ideals about universal love and expanded consciousness could coexist with exploitation, abuse, and the dangerous rejection of protective social structures. When Understanding Ruth Israel, who lived in the Love Family for 26 years, told me about discovering that a 12-year-old girl in the commune was pregnant and hearing the child say "my mom's going to be a grandmother… I'm going to be 13," it crystallized a hard truth: the hippies' insights about the nature of consciousness were often profound and scientifically prescient, but their attempts to translate these insights into social structures frequently failed catastrophically. The same people who intuited that "it's all vibrations, man" sometimes created communities where the most basic vibrations of human decency and responsibility broke down. Understanding both the visionary successes and the tragic failures of the counterculture is essential if we want to salvage its genuine wisdom while avoiding its most destructive mistakes.

As we explore these ideas, I invite you to maintain both an open mind and a critical eye. The questions we're addressing—about the nature of consciousness, its relationship to physical processes, and its distribution throughout the universe—are among the most profound we can ask. They touch not just on science and philosophy but on our deepest sense of who and what we are.

So tune in, but don't drop out entirely, not yet. And don't turn off your critical faculties. The science and other ideas I'm about to share may be as mind-expanding as any acid trip, but probably more verifiable.

Chapter One

What is This, *This* Consciousness?

I lost myself. Somewhere, somewhere deep. Everything I had ever thought I was, everything I had ever believed about the nature of reality simply… evaporated. *Poof. Whooooooosh!* And then, eons later, slowly, tentatively, consciousness reassembled itself from scattered fragments. *I* returned.

My journey on that inflatable bed in Los Angeles, surrounded by loving strangers and one good friend, lasted in the manifest world we inhabit about only forty-five minutes from start to finish, but opened many questions I'd been exploring my entire life, and am still exploring now. It wasn't "just a drug experience"—it seemed to be a direct encounter with Source itself, in its many-layered forms.

For forty years, I had been Tam—the skeptical philosopher, the former Army soldier, the idealistic lawyer, who approached these questions with careful academic distance. But something fundamental had shifted during those timeless moments when "Tam" had dissolved completely.

I had spent decades thinking about consciousness from the outside. But nothing—no amount of scientific training, no philosophical preparation—had readied me for the direct encounter with consciousness as the fundamental fabric of existence itself. The hippies, it seemed, had been onto something far more profound than their critics ever imagined.

As I reintegrated on the "launch pad," with arms outstretched and chest raised, coming back into my body, to me, I breathed deeply and rhythmically. I found myself trying to form words, wanting to express something to those around me. But I realized that words could not possibly do justice to the experience. *Why even try?* All I could do was hold my chest and gesticulate outwards, eyes still closed, tears of joy and bliss welling and falling down.

The only words I spoke during my journey were "no fear." I realized that I should have had no fear as I jumped into the abyss, and that humans are generally too consumed by fear. We need to learn to let go.

So, what is *this*? This moment, here now. My awareness. My consciousness. How did I come into being? Who are *you*? What are we all doing here in this strange reality? How is it possible that a relatively simple molecule can so quickly induce profound changes in our experience like it did for me in the moments I've described.

These questions lie at the heart of what philosopher David Chalmers famously dubbed "the hard problem of consciousness" in 1995. To understand why this problem is so hard, we need to distinguish it from what Chalmers called the "easy problems"—explaining specific mental functions like attention, memory, and behavior. While these "easy problems" remain scientifically challenging, they seem solvable using the tools of neuroscience, cognitive science, and psychology.

Think about it this way: imagine the most complete explanation of what happens in your brain when you see the color red. This explanation might detail how light of certain wavelengths hits your retina, how signals travel through your visual system, and which brain regions process this information. It might specify exactly which neurons fire and in what patterns. Yet somehow, this exhaustive physical description seems to leave something crucial out—the subjective experience of redness itself, what philosophers call "qualia." How does all that physical activity somehow create the experience of redness?

This explanatory gap between physical descriptions and subjective experience has troubled philosophers for centuries. René Descartes proposed that mind and matter are fundamentally different substances. This view has fallen out of scientific favor because it struggles to explain how a non-physical mind could interact with the physical brain. Modern materialist approaches that identify consciousness with brain activity face the opposite problem—they struggle to explain how electrochemical signals, no matter how complex, could give rise to the richness of conscious experience.

In recent decades, neuroscience has made remarkable progress in identifying the "neural correlates of consciousness" (NCCs)—the specific patterns of brain activity that correspond to conscious experiences. Researchers like Francis Crick and Christof Koch pioneered this approach, focusing on how neural synchrony, particularly in the gamma frequency range (30–100 Hz), seems to correlate with conscious perception (Koch has since backed away from this particular hypothesis but it still has some validity, albeit being just part of the picture of what's going on). When large groups of neurons fire together at certain frequencies, consciousness seems to be present. When this synchrony is disrupted, consciousness changes or disappears.

This research has given us valuable insights. We now know that consciousness seems to involve the coordinated activity of the thalamus (a central relay station in the brain) and the cerebral cortex (the wrinkled outer layer). We know that certain frequencies of brain waves correlate strongly with conscious states. And we know that disrupting this synchrony—through anesthesia, seizures, or deep sleep—profoundly alters or eliminates consciousness.

Yet these findings, significant as they are, primarily address the "easy problems" of consciousness—they tell us which neural processes, and the fields that they generate, correlate with and enable consciousness, but not why or how these processes give rise to subjective experience. As Christof Koch himself acknowledged in a 2019 article, no one has yet explained why any physical system, whether a brain or a computer, should give rise to felt experiences that are so different from physical systems.

This persistent gap has led some scientists and philosophers, including me, to reconsider an ancient idea called panpsychism—the view that consciousness is fundamental and everywhere, present in some form at every level of reality. In this view, consciousness doesn't suddenly emerge when matter reaches a certain level of complexity; it's always there, becoming more complex and rich as the physical structures that support it become more integrated. Where there is mind, there is matter; where there is matter, there is mind. They go together like black and white.

The panpsychist approach avoids the hard problem of explaining how consciousness emerges from non-conscious matter by suggesting that it doesn't emerge at all—it's a fundamental feature of reality. But this view requires that we also address what philosophers call the "combination problem"—how do these simpler forms of consciousness combine to form the highly complex, unified consciousness we experience?

This is where resonance enters the picture. When physical systems resonate in proximity—when they vibrate at the same or related frequencies—they can exchange energy and information more efficiently. Resonance creates a phase transition, allowing previously separate systems to function as a unified whole. In the context of consciousness, *shared resonance* is the mechanism by which micro-conscious entities combine into macro-conscious experiences. Saying "shared" resonance is admittedly a little redundant, since resonance is defined as a rhythm that is shared between two or more parts. Nevertheless, I like to use the phrase "shared resonance" to emphasize the shared aspects that are the core part of the workings of consciousness.

Consider what happens in your brain when you perceive a red apple. Light reflected from the apple stimulates your retina, sending signals to your visual cortex. Various specialized neurons process different aspects of the apple—its color, shape, location, and movement. But somehow, you don't experience these features separately; you have a unified perception of a red apple. This "binding problem"—how different features processed in different brain regions are integrated into a coherent perception—has long puzzled neuroscientists.

Neural synchrony, and the electromagnetic fields it produces, offers a potential solution. When neurons processing different aspects of the apple fire in synchrony—oscillating at the same frequency and phase—they effectively create a temporary network. It's as if they're all humming the same note, creating a chord that represents the unified apple. The electromagnetic field generated by this synchronized activity may itself be the substrate of your unified conscious perception—the "music" of your experience. (I explore this particular idea more fully in later chapters.)

This perspective aligns with what Walter Freeman, the late UC Berkeley professor and pioneer in brain research, observed in his studies on how rabbits process smell. Freeman found that when rabbits learned to recognize specific odors, their brains developed characteristic patterns of neural synchrony. These patterns weren't simply responses to chemical stimuli; they were shaped by the rabbits' learning and previous experiences. Freeman concluded that consciousness might reside not in the firing of individual neurons but in the dynamic patterns of the brain's electromagnetic field—the resonant "music" created by neurons playing together. This is a layer above the layer of neural firing and has far more information content than neural firing itself.

Similar insights have emerged from the work of Pascal Fries, whose "communication through coherence" hypothesis suggests that neural populations communicate effectively when their oscillations synchronize, creating windows of opportunity for information exchange. When two brain regions, and the fields they produce, oscillate in sync, signals sent from one region arrive at the other when it's most receptive, maximizing information transfer. It's like two people dancing together—when they move to the same rhythm, their movements flow smoothly together. And when they don't, there may be toes getting stepped on.

These findings have led some researchers, including Johnjoe McFadden and Susan Pockett, and myself, to propose various electromagnetic field theories of consciousness. These theories suggest that consciousness resides not in the firing of neurons *per se*, but in the electromagnetic fields generated by this activity. Since electromagnetic fields are unified and continuous, they naturally address the binding problem—different aspects of experience are integrated within the field itself, like different instruments in an orchestra creating a single piece of music. These fields also hold more information than neural firing itself. Far far more information. We'll explore in later chapters exactly how much more information can be contained in these fields.

General Resonance Theory, the theory I've developed with Jonathan Schooler, builds on these insights while going a step further. GRT proposes that consciousness exists across a continuum of complexity, from the most rudimentary forms associated with simple physical systems to the rich consciousness we experience as humans, and perhaps even at higher levels of organization. What determines the complexity of consciousness is the degree of resonant integration a system achieves—how effectively its components synchronize to create unified patterns.

In this framework, the "hard problem" splits into two parts: a metaphysical question about the fundamental nature of consciousness and its relation to matter, and a scientific question about how conscious entities combine. The metaphysical question about why consciousness exists can be addressed through a panpsychist framework where consciousness is fundamental to reality itself and is an inherent quality of matter. Consciousness isn't something that emerges from non-conscious matter, but is built-in to what we call matter as its interior aspect. In this view, consciousness and matter go hand in hand—they are the inside and outside of each real thing.

All matter is inherently experiential because mind and matter are two aspects of the same thing: what I've called at times "menter." Instead of trying to derive experience from non-experiential matter (which creates a hard-to-overcome explanatory gap in terms of explaining how something so different from unfeeling matter, consciousness, could somehow emerge spontaneously from said matter), this approach recognizes that the universe contains subjects of consciousness all the way down to its most fundamental constituents. While this perspective posits certain brute facts—that the stuff of the universe is inherently experiential—it provides an intellectually satisfying worldview that requires no miraculous emergence of consciousness from wholly non-conscious matter. It's always there in some form. From the lowliest specks of matter to the most grand.

But the "combination problem" of consciousness—which asks how these micro-conscious entities combine into macro-scale conscious systems like us—seems more scientifically tractable, more empirical. If shared resonance is the mechanism for this combination, we can investigate it empirically by studying how resonance operates at different levels of organization and how it facilitates information integration and cognition.

This approach doesn't eliminate the mystery of consciousness, but reframes it in a way that opens up new avenues for scientific investigation. Instead of asking how consciousness emerges from non-conscious matter, we ask how conscious entities at one level combine through resonance to create more complex conscious entities at higher levels.

This perspective has potentially profound implications. It starts by accepting that consciousness isn't an all-or-nothing phenomenon that suddenly appears at some threshold of complexity—which remains the most common view among neuroscientists and philosophers. Instead, consciousness is present in some form at all levels of reality, becoming richer and more complex as systems achieve greater resonant integration, and particularly through the process of biological evolution in animals.

Yes, this view aligns in key ways with the intuitions of many contemplative traditions that have long suggested consciousness permeates reality—intuitions often dismissed by modern science but perhaps worthy of reconsideration in light of new data.

The hippies were onto something when they talked about "cosmic consciousness" and universal vibrations. The universe is indeed composed of nothing but vibrating fields, resonating at different frequencies. Matter itself is a form of concentrated energy, oscillating in patterns we perceive as solid objects because of their relative constancy. Our conscious experience is another manifestation of these same fundamental dynamics—not something magical or separate from the physical world, but a natural expression of its inherent properties.

This doesn't mean we should uncritically embrace every aspect of New Age thinking. Many specific claims made by proponents of "cosmic consciousness" remain dubious. But the core intuition—that consciousness is fundamental and related to resonance—deserves serious scientific attention. And, yes, that resonance runs deeper than the manifest world we find ourselves in.

In 2019, I and my colleague and long-time friend and co-adventurer, Jonathan Schooler, published "The Easy Part of the Hard Problem: A Resonance Theory of Consciousness" in *Frontiers in Human Neuroscience*, which attempted to tackle key aspects of the classic hard problem. We had many collegial discussions about which part should be labeled the "easy" part and which the "hard" part—a distinction that was admittedly somewhat tongue-in-cheek given that both aspects remain profoundly challenging. I argued that the metaphysical question of how mind relates to matter constitutes the "easy part," which we addressed by adopting panpsychism as our starting point, while the "hard part" involved the specific mechanisms by which micro-conscious entities combine into macro-conscious ones—what philosophers call the "combination problem," as we've already seen. Jonathan thought it was the other way around and was more cautious about fully embracing panpsychism. Our paper proposed that synchronized oscillations—shared resonance—provide the mechanism for this combination into more complex conscious entities, allowing micro-conscious entities to integrate into macro-conscious wholes through phase transitions in information flow speed and bandwidth (later chapters will explain what this means!). While we disagreed on some philosophical foundations, we were united in seeing resonance as key to understanding how the many become one in conscious experience. And since then, our approach and our theory have… resonated widely with scholars and laypeople around the world.

In the next chapter, we'll dive deeper into the physics of waves and fields, examining how resonance operates at different scales of organization—from quantum systems to neural networks to social groups. We'll explore how modern physics has moved away from the view of matter as solid particles toward a view of reality as composed of vibrating fields. And we'll consider how this shift in our understanding of the physical world might inform our approach to the hard problem of consciousness.

The hard problem remains hard. But by reframing it in terms of resonance and integration rather than miraculous emergence, we might find that parts of it are more tractable than previously thought. The hippies may not have had the scientific vocabulary to articulate their intuitions precisely, but their fundamental insight—that vibration, resonance, and coherence are central to understanding consciousness—may prove to be one of their most enduring contributions to human knowledge.

Chapter Two

Waves, Fields, & Vibrations are the New Physics

During my Army years, from age 19 to 23, I felt starved for intellectual stimulation and longed to be back in college. I wrote in my journal during my second year in the Army, in a distinctively early 20s kind of tone: "This is a problem in my current environment. There aren't any intellectuals here. My life has always been future-oriented and so it remains. I hope to find the sought out intellectuals in the college environment, where else could they be?" What I did instead of being in college was read books—a lot. Among the hundreds of books I devoured during those years, about two a week, was Timothy Leary's book, *The Politics of Ecstasy*, a radical manifesto that challenged every aspect of the conventional reality I was ostensibly serving to protect.

Leary's central message—that psychedelics could be tools for deconditioning the mind from societal programming and accessing higher states of consciousness—resonated with an intuition I couldn't yet articulate. His famous directive to "turn on, tune in, drop out" wasn't just countercultural rebellion; it was, I began to suspect, a crude but effective protocol for altering the resonant frequency of consciousness itself.

This suspicion transformed into something approaching certainty several months later when, on a weekend trip from Oahu (where I was stationed at Schofield Barracks, the headquarters of the US Army's 25th Infantry Division), I ventured alone to Kauai, after a bad breakup from a short-lived but intense relationship with one of my commanding officers, a young captain who I won't name (yes, it was more than a little scandalous for me, an enlisted soldier, an E4 at the time, to be dating an officer, let alone an officer in my chain of command as she was), and hiked down to the bottom of Waimea Canyon with what experienced psychonauts would call a "heroic dose" of LSD. This is a term that Terence McKenna liked to use. McKenna was a more recent celebrity in the psychedelic world, dying in 2000, about four years later than Leary. He was an advocate of larger doses, in the right set and setting, in order to truly grasp the insights that these medicines and tools for consciousness exploration had to offer. For me, four tabs of acid was my heroic dose. That was the two tabs meant for her and the two meant for me. Two plus two equals four hits of mind-blowing insights.

I took the dose at the top of the canyon and by the time I got to the bottom about an hour later I was starting to trip hard. I found a beautiful pool of water and, after not seeing anyone else for the last half hour, stripped naked, and gave myself up to the experience. As the boundaries between myself and the towering red walls of the canyon began to dissolve, I experienced what can only be described as a radical shift in the frequency of my perception. The world hadn't changed, but the receiver—my consciousness—had been dramatically retuned.

Leary's opening essay in *The Politics of Ecstasy*—"The Seven Tongues of God"—struck me with particular force as I read it that night in my car at the parking lot above Waimea Canyon. It was as if Leary had written a guidebook to territory I had stumbled into, fallen into as I hiked down into that canyon—and then back up.

In this seminal essay, Leary articulated what I had intuited but couldn't name: that the human brain contains multiple "circuits" capable of producing different levels of consciousness. What mainstream science dismissed as hallucination, Leary framed as access to legitimate alternative realities—different "tongues" or languages through which the divine (or the fundamental nature of reality) could be experienced.

Leary's framework offered a map for understanding what had happened to me in that canyon. My usual perception—dominated by ego, verbal thinking, and the conventional boundaries between self and world—had temporarily given way to a different circuit of awareness. In that state, information normally filtered out by my everyday consciousness became accessible. The canyon walls weren't just red rocks; they were dynamic energy patterns in constant relationship with my own energy field. Time wasn't only a linear arrow but could also be a multidimensional medium that could expand, contract, and fold in upon itself.

Most revolutionary (to the Western mind) was Leary's suggestion that these altered states weren't aberrations or pathologies but potentially valuable tools for exploring consciousness. He challenged the assumption that ordinary waking consciousness represents the only legitimate way of perceiving reality. Instead, he proposed that different states of consciousness might reveal different aspects of a multifaceted reality—like different instruments in an orchestra playing the same cosmic symphony. I had *been there*, I knew these truths. I had experienced reality from a very different perspective and retained some of that perspective back in my "normal" state.

Reading Leary's essay was like finding intellectual validation for experiences that our mainstream culture had no vocabulary to describe. And yet, even as I resonated with his vision, I recognized that Leary's framework, revolutionary as it was for its time, remained limited by the scientific paradigms available to him in the 1960s. His "circuits" model, while useful as metaphor, didn't fully capture what modern physics now tells us about field-based reality.

What Leary described as discrete "circuits" of consciousness might better be understood as different patterns of resonance in the brain's electromagnetic field—different "tunings" that allow access to different aspects of information. The "seven tongues" he identified correspond to distinct resonant frequencies, each with its own characteristic pattern of information integration.

Looking back now, I see that Leary was reaching toward the field-based understanding of consciousness that contemporary neuroscience is only now beginning to formalize. His talk of "tuning in" wasn't just hippie metaphor but an intuitive grasp of what we now understand about neural synchrony and field resonance. When he urged us to "turn on, tune in, drop out," he was essentially describing a protocol for altering the resonant frequency of consciousness—"turning on" new circuits (activating different resonance patterns), "tuning in" to the information available at these frequencies (achieving coherent resonance), and "dropping out" of our culturally conditioned patterns of perception (releasing the dominant resonance patterns that maintain our conventional reality tunnels).

My Waimea Canyon experience had been exactly that—turning on, tuning in, dropping out. At least temporarily. The LSD had momentarily disrupted my brain's default resonance patterns, allowing new patterns to emerge. These new patterns permitted information exchange across neural networks that normally don't communicate, creating the expanded awareness characteristic of psychedelic states. Everything melted, not just visually, but conceptually, into each other.

The modern physics of fields and waves gives us a vocabulary that neither my canyon-shocked self nor Leary in the 1960s possessed. But the experience itself—that direct perception of reality as vibrating, interconnected fields rather than separate solid objects—has remained constant across generations. From mystics to hippies to quantum physicists, those who glimpse this underlying nature of reality often return with strikingly similar reports, even when they lack a common language to describe what they've witnessed.

What I experienced in Waimea Canyon wasn't just a hallucination, though mainstream science would classify it as such. It was a profound alteration in how my consciousness processed and integrated information. Colors intensified not because my visual system was malfunctioning, but because the usual filters had been temporarily bypassed. The sense of unity with my surroundings wasn't delusion but the temporary suspension of the boundaries my brain normally constructs between "self" and "world."

Years later, as I began studying physics and neuroscience formally, I realized that modern physics had been moving toward a similarly radical reconceptualization of reality—away from the Newtonian model of solid particles in empty space and toward a view of reality as composed of vibrating fields in dynamic interaction.

From Particles to Fields: The Quantum Revolution

Newton's physics gave us a world of solid objects—tiny billiard balls careening through empty space, interacting through direct contact or action at a distance. This mechanistic view shaped not just science but our entire conception of reality for centuries. It's still the implicit model most people carry in their heads: the world as a collection of solid things moving in time.

But this model began to crumble in the early twentieth century, first with the development of relativity theory and shortly after with the development of quantum mechanics. The discovery that light could behave both as a wave and a particle (the famous wave-particle duality) was just the beginning. As physicists probed deeper into the subatomic realm, they found that what we call "particles"—electrons, protons, quarks—aren't tiny solid objects at all. They're better understood as excitations in underlying quantum fields that permeate all of space.

Here's a simpler way to think about it: imagine the universe isn't made of tiny billiard balls floating in empty space, but instead is filled with invisible fields—including the magnetic fields you might have visualized in school by sprinkling iron filings on paper. These fields extend everywhere. What we call "particles" are just places where these fields are concentrated, and vibrating or excited, like ripples on a pond. An electron isn't a tiny sphere; it's a condensed vibration in the electron field that extends throughout the universe.

This shift from particles to fields represents one of the most profound conceptual revolutions in the history of physics. This field-based view of reality aligns remarkably well with certain mystical traditions that describe reality as a vibrating energy field. My father wasn't wrong when he talked about "vibrations"—he just lacked the mathematical or philosophical framework to express his intuitions precisely. When a quantum physicist speaks of reality as vibrating fields and a psychedelic voyager describes reality as vibrating energy, they may be approaching the same underlying truth from different angles.

Resonance: A Universal Phenomenon

Within this field-based reality, resonance emerges as a fundamental organizing principle. Resonance occurs when a system absorbs energy most efficiently because the frequency of the incoming energy matches the system's natural frequency.

The classic example is pushing a child on a swing—if you push at just the right moment (in phase with the swing's natural rhythm), small pushes can create large oscillations. Push at the wrong moment, and you disrupt the swing's motion instead. This simple playground physics reveals something profound about how things interact throughout nature.

But resonance isn't just about playground physics. It's a phenomenon that appears at every scale of organization. At the quantum level, particles can enter states of quantum coherence where their wave functions become synchronized, leading to phenomena like superconductivity, where electrons flow without resistance. This isn't science fiction—it's the principle behind MRI machines in hospitals worldwide.

In molecular biology, resonance stabilizes molecular structures and facilitates energy transfer in processes like photosynthesis. When plants capture sunlight, the energy doesn't simply travel randomly through the photosynthetic complex. It follows quantum coherent pathways—a kind of resonance that allows remarkably efficient energy transfer. Nature has been exploiting quantum resonance effects for billions of years.

In neural systems, resonance enables neurons to respond preferentially to inputs at certain frequencies. Your brain isn't just a collection of cells firing randomly; it's an orchestra of oscillating networks playing in complex harmony. Different brain states—from deep sleep to focused attention to creative insight—correspond to different resonant patterns across neural populations and the fields they produce.

In social systems, resonance manifests as the synchronization of behavior in groups. You've probably experienced how audiences begin applauding in sync, or how your footsteps naturally align with your walking companion's. These spontaneous synchronizations happen without conscious effort—they're emergent properties of resonant systems.

What makes resonance so interesting in the context of consciousness is that it provides a mechanism for how separate entities can become integrated into functional wholes. When two oscillating systems come into resonance, they begin to exchange energy and information more efficiently. They become, in a very real sense, entangled—their behavior can no longer be fully understood by looking at each system in isolation.

Consider the smartphone in your pocket. When you make a call, your phone converts your voice into electromagnetic waves that precisely match the receiving frequency of cell towers. If these frequencies didn't resonate exactly, your conversation would be lost in the sea of electromagnetic noise filling the air around us. The principle is the same in your brain—neurons that resonate at precise frequencies communicate effectively, while those oscillating out of sync remain functionally disconnected.

This principle of integration through resonance offers a potential solution to one of the central puzzles of consciousness—how the activities of billions of neurons, each with their own individual dynamics, combine to create the unified field of conscious experience. If consciousness is fundamentally a resonance phenomenon, then the question becomes not how consciousness emerges from non-conscious matter, but how conscious entities at one level combine through resonance to create more complex conscious entities at higher levels.

Electromagnetic Fields and Consciousness

Of particular interest for understanding consciousness are electromagnetic fields—the same fields that give rise to visible light, radio waves, and the forces that hold atoms together in molecules. The brain's electrical activity, measurable through techniques like EEG (electroencephalography) and MEG (magnetoencephalography), generates electromagnetic fields that permeate and extend beyond neural tissue.

These fields aren't just byproducts of neural firing—epiphenomena with no causal role. They're physical entities that can influence the very neurons that generate them, creating feedback loops of mutual causation. This was demonstrated dramatically in experiments where weak external electric fields, similar in strength to those generated naturally by the brain, were shown to entrain neural firing patterns.

Think of it this way: when neurons fire, they create tiny electrical currents that contribute to an overall electromagnetic field. This field, in turn, can influence when those same neurons will fire next. It's a two-way street, not a one-way causation. My mother's spinning wheel provides a helpful analogy here—the momentum of the wheel influences how easily she can continue pedaling, while her pedaling sustains the wheel's rotation. Neither is simply causing the other; they're engaged in continuous circular causation.

And it's not just that the brain's electrical fields influence firing, it's much more complex than that. The brain's electrical fields have their own causal structure that can carry far more information than is contained in the neural firing. In the same way that a tree's trunk and larger branches support and grow the smaller branches, twigs and leaves, so the brain's neural structure "grows" the brain's electric fields. And just as there is far more complex structure, and thus information, in the tree's smaller branches, twigs, and leaves, so there is far more complex structure, and thus information, in the brain's vastly complex and intricately timed field structure.

Several scientists have proposed that the brain's electromagnetic field might be the physical basis of consciousness itself. Johnjoe McFadden's Conscious Electromagnetic Information (CEMI) field theory suggests that since this field is unified and continuous by its very nature, it provides a natural solution to the binding problem —how different aspects of experience are integrated into a coherent whole.

Similarly, Susan Pockett has argued that consciousness is identical with certain spatiotemporal patterns in the brain's electromagnetic field. She suggests that if we could recreate these precise patterns artificially, we would recreate the associated conscious experiences—a testable prediction that distinguishes field theories from competing accounts. My own GRT approach would agree with these aspects of both McFadden's and Pockett's theories.

These approaches share a recognition that consciousness seems to involve the integration of information across space and time—exactly what resonating fields excel at facilitating. When neurons fire in synchrony, their individual electromagnetic contributions combine constructively, creating stronger fields that can influence neural activity over greater distances. This enables the formation of transient functional networks that seem to correspond to specific conscious contents.

Our brains create unified conscious experiences through patterns of electromagnetic synchrony that bind disparate neural processes into coherent wholes.

Fields Within Fields: The Nested Structure of Reality

One of the most profound insights from modern physics is that reality has a nested, hierarchical structure—fields within fields within fields. Elementary particles are excitations in quantum fields. Atoms are organized systems of these particle-waves.

Molecules are structured collections of atoms. Cells are complex molecular systems. Organisms are integrated cellular networks. Ecosystems are interconnected webs of organisms. And so on.

At each level of organization, new properties emerge that aren't readily predictable from the properties of the constituent parts. The wetness of water isn't contained in individual H_2O molecules; it emerges from their collective interactions. The metabolism of a cell isn't present in its component molecules; it arises from their integrated functioning as a system. The rudimentary consciousness associated with simple physical systems becomes progressively more complex and integrated as those systems themselves become more integrated.

I witnessed this principle in my mother's handcrafts—a simple stitch repeated thousands of times created a sweater with properties the individual stitches didn't possess. The sweater could keep me warm, could communicate her care through its design, could connect to cultural traditions through its patterns. None of these properties existed at the level of individual stitches, yet they weren't mysterious additions; they emerged naturally from the organization of the whole.

Similarly, the rich conscious experiences we enjoy as humans aren't contained in individual neurons, nor do they require the addition of some non-physical element. They arise from the specific patterns of resonant integration that our complex brains support—patterns that create new properties while still being fully grounded in physical processes.

This perspective aligns with what philosopher Alfred North Whitehead called "process philosophy," where reality is seen not as a collection of static objects but as a dynamic process of becoming. In Whitehead's view, every actual entity—from elementary particles to human beings—has both a physical and a mental aspect. The mental aspect is what we might call consciousness or experience; it's the inside view of what appears from the outside as physical behavior.

Whitehead's process philosophy anticipated many aspects of contemporary physics, including the shift from particles to fields and the recognition of the dynamic, relational nature of reality. It offers a metaphysical framework compatible with the field-based ontology of modern physics and with the resonance-based approach to consciousness I'm advocating.

Testing the Field View: Empirical Approaches

The field-based view of consciousness isn't just philosophical speculation; it generates testable predictions. If consciousness resides in electromagnetic fields, then manipulating these fields should directly affect conscious experience. And indeed, techniques like transcranial magnetic stimulation (TMS) and transcranial direct current stimulation (tDCS), which alter the brain's electromagnetic fields, can profoundly affect perception, cognition, and even mood.

More specifically, if neural synchrony—the coordinated oscillation of neural populations—is crucial for consciousness, then disrupting this synchrony should disrupt consciousness. This prediction is borne out by research on anesthesia, which appears to work in part by disrupting the synchronization of neural activity, particularly in the gamma frequency range (30–100 Hz) associated with conscious awareness.

Conversely, enhancing neural synchrony through techniques like neurofeedback or rhythmic sensory stimulation should enhance aspects of conscious experience. Preliminary research suggests this may indeed be the case, with studies showing that inducing gamma synchrony can enhance cognitive performance and even facilitate lucid dreaming—a state where the dreamer becomes aware they're dreaming while continuing to dream.

Perhaps most intriguingly, if consciousness is a field phenomenon that exists across a continuum of complexity, we might expect to find evidence of consciousness-like properties in systems usually considered non-conscious—simple organisms, individual cells, or even complex molecular systems. And while this remains a frontier area, there is intriguing evidence that single-celled organisms like paramecia can learn and remember, that plants communicate and respond adaptively to their environments, and that certain molecular systems display complex, coordinated behaviors that resemble intentionality.

These observations don't prove that such systems are conscious in the rich sense that humans are. But they suggest that the line between conscious and non-conscious systems may be less sharp than commonly assumed—exactly what we would expect if consciousness exists on a continuum rather than appearing suddenly at some threshold.

Back to the Canyon: Integrating the Personal and the Scientific

As I sat at the bottom of Waimea Canyon that day in 1994, watching my sense of self temporarily dissolve into the red rock walls, I had no vocabulary to articulate what was happening at the level of field dynamics or information processing. But I did have an overwhelming sense that I was experiencing, in some direct way, the underlying unity of reality—that the boundaries between myself and the world, usually so stark, were revealed as constructions of my nervous system rather than fundamental divisions in reality itself.

Timothy Leary had written about this exact experience, which he categorized as "Stage Five" in his eight-stage model of psychedelic consciousness—"merging with external objects" where "all boundaries melt," and the usual distinctions between self and other temporarily collapse. What Leary described phenomenologically, modern neuroscience might characterize as a profound shift in the brain's default mode network activity, disrupting the usual processes that maintain the sense of a separate self.

But neither Leary's phenomenological account nor the neuroscientific explanation by itself captures the full significance of such experiences. By integrating them with field-based physics, we gain a more comprehensive understanding: the dissolution of boundaries experienced during psychedelic states may represent a temporary reconfiguration of the electromagnetic fields associated with consciousness, allowing for greater coherence between the individual's field and the fields of the surrounding environment.

This isn't to suggest that psychedelics are necessary or sufficient for understanding consciousness scientifically. But they do offer a valuable first-person perspective on how consciousness can be radically reconfigured—a perspective that complements third-person methodologies like neuroimaging and electrophysiology. By integrating these perspectives with the field-based ontology of modern physics, we can develop a more complete picture of consciousness and its place in the physical world. Third-person science meets first-person science.

The New Physics of Consciousness

The shift from a particle-based to a field-based physics parallels a potential shift in our understanding of consciousness—from localizing it in the firing of specific neurons to, instead, recognizing it as a field phenomenon that integrates information across the brain and perhaps beyond to the body and its environment too. This shift doesn't necessarily negate the importance of neurons and their connections; rather, it places them in a broader context, recognizing that the electromagnetic fields they generate may be as causally significant as the electrochemical processes that receive the lion's share of attention in standard neuroscience.

The hippies were right about waves and fields being fundamental to reality, if not always about the specific mechanisms involved. Their intuition that consciousness involves resonance with larger fields—what they might have called "cosmic consciousness" or "universal vibrations"—finds a surprising echo in the field-based physics that has replaced the mechanistic Newtonian worldview.

This doesn't mean we should uncritically embrace every New Age claim about consciousness and energy fields. But it does suggest that the basic intuition of reality as vibrating fields in resonant interaction may be closer to the mark than the materialist view that dominated science throughout much of the modern era until the twentieth century.

In the next chapter, we'll explore how these insights from physics connect with findings from neuroscience, examining the specific mechanisms by which resonance might operate in neural systems to create the integrated, unified field of consciousness we each experience. We'll see how the brain's layered structure of oscillating networks—from the fastest gamma waves to the slower theta, alpha, and delta rhythms—creates a nested hierarchy of resonance that may correspond to the nested structure of conscious experience itself.

But for now, I want to emphasize the profound paradigm shift represented by the move from particles to fields. This shift invites us to rethink not just our understanding of consciousness but our entire conception of reality. If the fundamental nature of reality is vibrational, resonant, and field-based, then the qualitative aspects of experience—the redness of red, the sweetness of sweet, the emotional texture of joy or sorrow—may not be additions to the physical world but expressions of its intrinsic nature, accessed through the particular resonant structures of our nervous systems. They were there all along.

The day I spent in Waimea Canyon, psychedelically dismantled, pulverized, and reconstituted, forever changed how I think about consciousness and my place in the universe. It set me on a path that eventually led from the military to academia, from Timothy Leary to quantum field theory, from my childhood in the commune to the cutting edge of consciousness science. The journey has been neither straight nor easy, but it has convinced me of one thing: the hippies were right about reality being fundamentally vibrational. And modern physics, properly understood, says essentially the same thing—just with more precise mathematics and fewer tie-dyed t-shirts (though I still unabashedly love tie-dye).

Chapter Three

The Ripples Where Science & Philosophy Meet

I am a Wattsist. A what-ist? A Wattsist, after Alan Watts, the British-American philosopher and public intellectual.

During my long evenings at Schofield Barracks in Hawaii, during my military years, when I should have been polishing my boots or preparing for morning inspection, I found myself repeatedly drawn to the Borders bookstore near Pearl Harbor. It was there that I discovered Alan Watts—a British-American philosopher whose words would provide solace and understanding throughout my life's journey. His books became my intellectual companions during those years of military service—an unlikely pairing that reflected the contradictions of my own journey.

Watts was a British expat who spent most of his life in the U.S., with his later years in Marin County and the Bay Area. His worldview and spiritual perspectives were shaped largely by Eastern traditions—in particular Zen Buddhism, Taoism, and Vedanta Hinduism—but also by his Christian upbringing and training. He was an Episcopal priest for six years, serving at Northwestern University in Illinois, until he left the church after he and his first wife messily separated.

What drew me to Watts was his remarkable ability to bridge Eastern and Western thought, translating profound philosophical concepts into accessible language without diluting their power. In books like *The Book: On the Taboo Against Knowing Who You Are* and *The Way of Zen*, Watts articulated a vision of reality that (slowly but steadily) resonated deeply with what I would later explore through scientific frameworks—that beneath the appearance of separation lies a fundamental unity, that consciousness may be more fundamental than we typically assume, and that our ordinary perception of reality as composed of separate "things" might be a kind of collective hallucination. Watts wrote *The Book...* for his son, Mark Watts (who I would eventually meet many years later), as a guide to essential spirituality. Apparently Mark was about 17 when Alan wrote the book for him, and I wasn't much older when I first encountered this remarkable volume.

Reading these works in my barracks, surrounded by the rigid hierarchy and mechanistic thinking of military life, created a cognitive dissonance that ultimately proved productive. I found myself mentally toggling between the orderly, pseudo-Newtonian world of military protocols and the more fluid, interconnected reality that Watts was describing. This tension became a crucible for my own developing thoughts about consciousness and reality.

Watts' constant themes—the perils of language tricking us into mistaken views about ourselves and reality, the pervasiveness of mind throughout reality, and most provocatively, the recognition that each of us is, in a profound sense, the entire universe experiencing itself from a particular viewpoint—would become touchstones for my own developing thoughts about consciousness and its place in the cosmos.

In his profound little book, *The Book...*, Watts writes:

> Because there is nothing outside God, he has no one but himself to play with. But he gets over this difficulty by pretending that he is not himself. This is his way of hiding from himself. He pretends that he is you and I and all the people in the world, all the animals, all the plants, all the rocks, and all the stars. In this way he has strange and wonderful adventures, some of which are terrible and frightening. But these are just like bad dreams, for when he wakes up they will disappear.

This playful yet profound metaphor—that the universe is essentially God playing hide-and-seek with itself—captured something I had intuited but struggled to articulate. It suggested that beneath the apparent multiplicity of existence lies a fundamental unity, a unity that each of us participates in fully rather than partially. Our normal life is a forgetting of this fundamental unity. Spiritual unity and practice is our way to remember.

I came to appreciate Watts not as a perfect sage—he was no angel, struggling with alcoholism and being, by his own admission, a less than ideal father and husband—but as a wonderfully eloquent voice articulating insights that bridged ancient wisdom and modern understanding. He described himself (and all of us too) as having an "irreducible rascality," acknowledging the contradictions that lived within him as they live within all of us. (A later chapter delves into this shadow side of the hippie movement.)

Watts wasn't a scientist, and some, including my own father, liked to dismiss his philosophical musings as insufficiently rigorous. But what makes his work remain so valuable is precisely its ability to translate between different ways of knowing—to find the resonances between mystical insight and scientific discovery, between direct experience and conceptual understanding.

As I've developed my own philosophical and scientific approach to consciousness through General Resonance Theory, I've found myself returning again and again to Watts' insights as touchstones—not as scientific hypotheses to be tested directly, but as articulations of experiential truths that scientific theories must ultimately accommodate if they are to provide a complete account of consciousness. I regularly return to his books and recorded talks to remind myself of his insights. His rich British voice is comforting. And his truths always profound and beautiful.

Watts' influence on my thinking represents a broader pattern—the ripples where science and philosophy meet, where different approaches to understanding consciousness can inform and enrich each other rather than existing in isolation or opposition. This integration of perspectives becomes particularly valuable when considering consciousness, which by its nature spans subjective experience and objective measurement, first-person and third-person perspectives.

Shortly after leaving the Army at 23, I had an experience that would eventually force me to confront this question with more focus. I was back in Seattle, spending time with my father and brother. My older brother had lived in Seattle for many years and he and I had enjoyed a relatively close bond over the years even though we were always quite different. We'd spent the evening in the University District where he lived, hopping between coffee shops, engaging in those meandering conversations with friends and strangers that seem to define the Seattle experience. When we returned to my brother's apartment, my father offered us a joint, and we all smoked together—one of those rare moments of uncomplicated connection between us.

The cannabis was potent, and we all retreated to bed shortly afterward, with me settling onto my brother's couch, fully "stoned." As I lay there feeling the pleasant body high that cannabis typically induces, something unexpected began to happen. What started as waves of physical relaxation transformed into pulsating ripples of bliss coursing up and down my body. This sensation intensified until I felt enveloped in a tube of white light. The experience crescendoed into what I can only call a "soulgasm" —a full-body, full-consciousness explosion of ecstasy that felt unmistakably like a classic mystical "seeing God" experience.

This wasn't anything like the typical effects of cannabis that I'd experienced before. It wasn't like my acid trips either, which had been more visual and cognitive in nature. This was something entirely different—direct, visceral, and unambiguously spiritual in quality. And it has never happened again to me, despite many subsequent cannabis experiences.

For years, I filed this anomalous experience away as an unexplainable aberration, a strange neurological hiccup perhaps, certainly nothing to build a whole worldview around. My scientific leanings urged me to discount it as meaningless, just a peculiar interaction between the cannabis and my brain chemistry that night. But the memory remained vibrant, refusing to fade or be rationalized away.

More than a decade later, after years of studying science and philosophy, and then attending law school at UCLA and becoming a lawyer (somewhat strangely), I came to recognize this experience for what it was: my first profound encounter with something that transcended my hardcore materialist framework. It was qualitatively different than my acid journeys had been. Even though those journeys expanded my awareness, even displaced it and thus allowed a kind of triangulation for examining reality and my place in it, I cannot describe my acid journeys as any kind of "seeing god" experience. This cannabis journey was a data point that my model of reality at that time couldn't accommodate—a direct experiential challenge to the notion that consciousness is merely an epiphenomenon of brain activity with no deeper significance, or that there is no deeper reality.

The experience didn't convert me to any religious or spiritual tradition. I remained committed to a broadly physicalist framework, but my version of physicalism had to expand dramatically to accommodate experiences like this. I gradually moved toward what philosophers call panpsychism—the view that consciousness or mind is an intrinsic feature of physical reality, present wherever matter exists, though in infinitely varying degrees of complexity and intensity, prodded by my reading various panpsychist thinkers during my Army years of active reading.

In this tentative new worldview, what happened on my brother's couch wasn't a violation of physical law or an intrusion from some supernatural realm. It was a momentary lifting of the veil that normally prevents us from directly experiencing the source consciousness that undergirds all of physical reality—a brief alignment of my individual consciousness with broader currents of awareness that are always present but rarely accessible to our everyday perception.

This question led me to further explore the intersections of Eastern philosophy and Western science—a borderland where the most profound insights about consciousness seemed to emerge. I began to see striking parallels between Watts' articulation of non-dualism and the field-based ontology of modern physics I was studying. Both suggested that the fundamental nature of reality was not discrete objects but interconnected processes—fields. Both challenged the conventional boundaries we draw between self and world. Both pointed toward a view of consciousness as intrinsic to nature rather than an anomalous emergence at some arbitrary midpoint of evolution.

This convergence between ancient wisdom traditions and cutting-edge science isn't coincidental. It reflects a deeper unity that manifests when we approach consciousness without the biases of scientific materialism or religious dogmatism. In this chapter, we'll explore this fertile meeting ground, examining how insights from philosophy, phenomenology, and contemplative traditions complement and enhance the scientific approach to consciousness I've been developing.

The Limits of Reductionism

Modern science has achieved remarkable success through reductionism—the approach of breaking complex systems down into their constituent parts to understand how they work. This method has given us everything from antibiotics to smartphones, from the periodic table to the human genome. But when it comes to consciousness, reductionism runs into fundamental limitations.

The problem isn't that reductionism is wrong—it's just incomplete. Analyzing the brain into neurons, neurons into synapses, synapses into molecules, and molecules into atoms and fields provides valuable insights. It makes complex systems easier to study and understand. It's been an invaluable tool in the development of modern science. But this bottom-up approach struggles to account for the emergent and relational properties that arise from complex systems—properties like life, mind, and consciousness that seem to be more than the sum of their parts.

Another limitation of traditional scientific approaches to consciousness is their exclusive reliance on third-person, objective methodologies. While invaluable, these methods inevitably miss something essential about consciousness—its inherently subjective, first-person nature. As Thomas Nagel famously argued in his paper "What Is It Like to Be a Bat?", no amount of objective data about bat physiology can tell us what it feels like to navigate the world through echolocation, if we're not already a bat and just *know* what that feels like.

This recognition has led to a renewed interest in phenomenology—the systematic study of experience from the first-person perspective. Pioneered by philosophers like Edmund Husserl, Maurice Merleau-Ponty, and Francisco Varela, phenomenology offers tools for investigating consciousness that complement the objective methods of neuroscience.

Varela, together with Evan Thompson and Eleanor Rosch, developed an approach called "neurophenomenology" that integrates phenomenological investigations with neuroscientific research. This approach doesn't privilege either first-person or third-person perspectives but recognizes both as essential for a complete understanding of consciousness.

From a resonance perspective, this integration makes perfect sense. The patterns of neural synchrony observable through techniques like EEG correspond to specific phenomenological states—what it feels like from the inside. These aren't two separate phenomena, but are instead two perspectives on the same underlying process—two sides of the same coin.

This dual-aspect approach helps resolve apparent contradictions between scientific and experiential accounts of consciousness. For instance, the unified field of conscious experience might seem at odds with the distributed, modular nature of brain processing. But if consciousness is understood as a field phenomenon, this apparent contradiction dissolves. The distributed processing in neural networks generates a unified electromagnetic field, which corresponds phenomenologically to our unified field of experience.

Contemplative Traditions: Laboratories of Consciousness

For thousands of years, contemplative traditions across cultures have developed sophisticated methods for investigating consciousness directly. Meditation, yoga, and other contemplative practices can be understood as rigorous first-person methodologies for studying the nature of mind—as laboratories of consciousness.

These traditions haven't just developed techniques; they've produced detailed maps of consciousness and theories about its nature that often align surprisingly well with contemporary scientific findings. For instance, Buddhist psychology's analysis of consciousness into discrete moments coheres with the modern understanding of neural processing as occurring in rhythmic cycles. The Yogic concept of consciousness as a field (*citta*, pronounced "cheeta") that can be modified by different patterns (*vrittis*) parallels field-based approaches to understanding consciousness.

What's particularly valuable about these traditions is their emphasis on direct experience rather than theoretical speculation. As the Zen saying goes, "Don't mistake the finger pointing at the moon for the moon itself." Contemplative practices invite us to look directly at consciousness rather than merely theorizing about it.

Neuroscientist Richard Davidson, founder of the Center for Healthy Minds at the University of Wisconsin-Madison, has pioneered research on the effects of meditation on the brain. His studies have shown that long-term meditation practice leads to measurable changes in brain function and structure, including enhanced neural synchrony in the gamma frequency range associated with conscious awareness.

Similarly, the work of Judson Brewer at Brown University has shown how meditation affects the brain's default mode network—a set of interconnected regions active when the mind is wandering and strongly associated with self-referential processing. Meditation decreases activity in this network, correlating with the phenomenological experience of reduced self-focus that meditators report.

These findings suggest that contemplative practices aren't just subjectively transformative but objectively measurable in their effects on the brain's resonant patterns. They offer a bridge between first-person and third-person approaches to consciousness, validating insights from ancient wisdom traditions through the lens of modern neuroscience.

Panpsychism: An Ancient Idea with Modern Relevance

The view that consciousness is fundamental to reality rather than an emergent property of complex brains—what philosophers call panpsychism—has ancient roots. We've touched on this view already a little, but let's dive a bit deeper now. This idea appears in various forms in Greek philosophy (particularly in the work of Thales and the Stoics), in indigenous worldviews worldwide, and also in Eastern philosophical traditions like Vedanta and certain schools of Buddhism.

For much of the modern scientific era, panpsychism was dismissed as unscientific—a relic of pre-scientific thinking, and often equated with *animism*, the view that all of nature is alive and ensouled to some degree. But in recent decades, panpsychism has experienced a remarkable resurgence among philosophers of mind and also among a growing number of scientists. This revival isn't a retreat to woolly mysticism but is, instead, a recognition that panpsychism might offer elegant solutions to persistent problems in explaining consciousness that avoid the pitfalls facing materialism and dualism.

Philosopher Galen Strawson has argued that panpsychism is actually more compatible with a thoroughgoing materialism than the emergence of consciousness from entirely non-conscious matter, a kind of emergence he considers "magical." If consciousness is real and physical (as materialism asserts), and if it can't emerge from completely non-conscious processes (as Strawson argues), then some form of consciousness must be fundamental to physical reality.

Similar arguments have been advanced by philosophers like David Chalmers and Philip Goff, and even by neuroscientists like Christof Koch, who has embraced a scientific form of panpsychism called "integrated information theory" (IIT) developed by Giulio Tononi, Koch, and others.

What's particularly interesting about modern scientific panpsychism is that it doesn't invoke supernatural entities or forces. It suggests that consciousness or proto-consciousness is an intrinsic aspect of the physical world, just like mass, charge, or spin. This aligns with the field-based ontology of modern physics, where reality consists fundamentally of fields with both extrinsic (observable) and intrinsic (unobservable) properties.

The Resonance Between Ancient Wisdom and Modern Science

The parallels between ancient wisdom traditions and cutting-edge science aren't superficial. They reflect a deeper resonance between insights arrived at through contemplative inquiry and those discovered through scientific investigation. Both point toward a view of consciousness as:

1. Field-like rather than particle-like;
2. Process-based rather than substance-based;
3. Relational rather than isolated;
4. Fundamental rather than emergent;
5. Unified yet differentiated.

This convergence doesn't mean that ancient mystics somehow anticipated quantum field theory or modern neuroscience. I'm not making that argument. Rather, it suggests that when humans investigate consciousness carefully—whether through contemplative practice or scientific inquiry—they tend to discover similar patterns and principles because both approaches are glimpsing various aspects of the underlying reality that we all live in and grow out of.

Alan Watts understood this when he wrote: "The mystical experience is nothing more than becoming aware of your own inmost energies, but it seems like something more than that only because we have been taught to find this sort of experience very peculiar, very unusual." From a resonance perspective, the mystical experience might be understood as a direct apprehension of the fundamental field-like nature of consciousness—an experience of the resonant unity that underlies the apparent separation between self and world.

General Resonance Theory: Bridging Science and Philosophy

From this diverse and unexpected backdrop, I began to develop my own theory of consciousness around 2010, after being invited to join Professor Jonathan Schooler's lab at UC Santa Barbara. We had met socially through a mutual friend, and had quickly began discussing consciousness and philosophy. I foolishly asked him over Red Bull and vodka: "Are you a grad student?" He replied, a little offended, "I'm a professor." Indeed, he is now a "distinguished professor" at UC Santa Barbara, which is the highest rank for professors in the UC system. We remain good friends and continue our collaboration and adventures of various sorts that began over fifteen years ago.

General Resonance Theory (GRT) offers a framework that bridges these scientific and philosophical approaches to consciousness. By identifying shared resonance as the mechanism by which conscious entities combine to form more complex conscious systems, GRT addresses both the scientific question of how consciousness is structured and the philosophical questions of how our consciousness relates to physical reality and our role in the universe.

From a GRT perspective, the insights of contemplative traditions aren't opposed to scientific understanding; they're complementary. When a meditator or psychonaut experiences the dissolution of the boundary between self and world, they may be directly apprehending the field-like nature of consciousness that scientific instruments detect as synchronized neural oscillations.

Similarly, when physicists describe reality in terms of vibrating fields rather than solid particles, they're approaching from a different angle the same insight that mystics express when they speak of the universe as a cosmic dance of energy or vibration.

This doesn't mean we should uncritically accept every claim from every wisdom tradition. Both scientific and contemplative approaches require rigorous evaluation of evidence, albeit different kinds of evidence. But it does suggest that dismissing millennia of sophisticated first-person investigation of consciousness would be as misguided as dismissing the third-person findings of neuroscience.

My own journey has convinced me that neither Eastern nor Western approaches to consciousness are complete on their own. Eastern traditions often excel at investigating consciousness from the inside, developing sophisticated maps of subjective experience. Western science excels at investigating the physical correlates of consciousness, developing precise tools for measuring brain activity. Both are essential for a complete understanding.

When I read Alan Watts in my army barracks three decades ago, I couldn't have articulated how his insights would eventually connect with the scientific study of consciousness. But looking back, I can see that Watts was pointing toward exactly the kind of integrated approach I'm advocating now—one that takes seriously both the scientific evidence and the direct experience of consciousness. Overcoming this taboo requires integrating multiple ways of knowing—combining the precision of science with the depth of contemplative inquiry, the objectivity of third-person methods with the immediacy of first-person experience.

This integration is what General Resonance Theory attempts to provide—a framework that honors both the materiality of consciousness (its basis in physical processes) and its irreducible subjectivity (its existence as felt experience). By identifying resonance as the key mechanism for the combination and integration of consciousness, GRT offers a path forward that transcends the historical divide between scientific and philosophical approaches.

The hippies were right when they intuited the fundamental unity of consciousness and cosmos, even if their articulation of this insight sometimes lacked scientific precision. And modern scientists are right to insist on empirical evidence and testable hypotheses, even if their methodologies sometimes exclude crucial aspects of conscious experience.

In the resonance between these approaches—scientific and contemplative, Western and Eastern, third-person and first-person—we may find a more complete understanding of consciousness than either could provide alone. As Watts himself wrote, with characteristic insight: "The world is not a collection of separate things but a pattern of relationships." This relational, process-oriented view of reality aligns perfectly with the field-based physics and resonance-centered account of consciousness I've been developing.

In the next chapter, we'll delve deeper into the scientific evidence for resonance-based consciousness, examining the specific mechanisms by which electromagnetic fields in the brain might support conscious experience. But we'll do so with the understanding that this third-person perspective is just one side of a coin whose other side is the direct, first-person experience of consciousness itself—an experience that contemplative traditions have been investigating for millennia.

The marriage of these approaches offers our best hope for addressing what David Chalmers called "the hard problem" of consciousness—explaining how physical processes give rise to subjective experience. By recognizing that physical processes themselves may have intrinsic, experiential aspects that science cannot directly observe but contemplative practices can access, we open the possibility of a more complete and integrated understanding of consciousness and its place in nature.

Chapter Four

Why the Hippies Became Hippies

In the summer of 1967, as an estimated 100,000 young people converged on San Francisco's Haight-Ashbury district for what the media dubbed the "Summer of Love," mainstream America watched in bewilderment. How had an entire generation seemingly rejected everything their parents had worked so hard to build? What had transformed clean-cut college students into long-haired, psychedelic-using, authority-questioning "freaks" who insisted on talking about peace, free love, going back to the land, and cosmic consciousness?

The hippie movement didn't emerge in a vacuum. It was the culmination of multiple converging forces that created the proverbial perfect storm of cultural transformation in America. Understanding these forces helps explain not just why the hippies became hippies, but why their insights about consciousness, interconnection, and alternative ways of living continue to resonate today—and why communities like the Love Family, where I would spend part of my childhood, seemed to offer such compelling alternatives to mainstream society.

The Vietnam Crucible: War, Draft, and Moral Crisis

Perhaps no single factor radicalized American youth more than the Vietnam War and the military draft that made their participation in it mandatory. Being forced into the military to go and kill brown people half a world away, for poorly defined reasons, didn't sit well with much of American youth—or their parents. As Jay Stevens documents in *Storming Heaven*, the war served as a catalyst that transformed what might have remained a small countercultural movement into a mass phenomenon affecting millions of young Americans.

Photo: My brother, Guido, Understanding Israel (the woman on the left) and I in the Love Family circa 1983 (I'm on the far left and Guido has a head band on and a duck leg in one hand).

The draft lottery system, instituted in 1969, meant that every young American man faced the very real possibility of being forced to fight in a war that increasingly seemed as unwinnable as it was morally questionable. College deferments created a temporary haven for some, but graduation loomed as a potential death sentence for many. This wasn't abstract political theory—it was an immediate existential crisis that forced an entire generation to confront fundamental questions about authority, morality, and the meaning of citizenship.

Beyond the immediate threat of the draft, the Vietnam War created a profound moral crisis that radicalized an entire generation. Television brought the horrors of war directly into American living rooms—images of napalmed villages, body counts reported like sports scores, the My Lai massacre revealing the profound depths of military brutality by targeting innocents with horrific weapons. Young Americans were being forced by their government to travel halfway around the world to kill and be killed by people who had never threatened them, poor peasants fighting for their independence just as Americans had fought the British two centuries earlier. The moral revulsion was visceral and widespread. Many young men faced an impossible choice: comply with a system that demanded they become instruments of imperial violence against innocent people, or resist and face imprisonment. This moral clarity about American empire and its violence created a generation that questioned not just specific policies but the entire system of power that could demand such participation in what many saw as genocide. The draft wasn't just threatening their lives—it was threatening their souls.

My father's experience illustrates the crucial difference between the American and British contexts. While American young men faced the draft, British youth like my father were free to explore alternative consciousness and countercultural values without the immediate threat of forced military service. The UK's non-involvement in Vietnam meant that British hippies could focus more purely on consciousness expansion, spiritual seeking, and cultural experimentation without the life-or-death urgency that was a significant part of the American resistance.

This difference helps explain why the hippie movement was more intense and politically charged in America than the UK and Europe. For American youth, consciousness expansion and political resistance became inseparable. The same questioning of authority that led young people to experiment with psychedelics naturally extended to questioning the authority that demanded they kill and die in Southeast Asia.

Stevens notes that many of the early psychedelic researchers, including Timothy Leary, were initially apolitical. But the war changed that, as it became impossible to separate consciousness expansion from political resistance. The phrase "Hey, hey, LBJ, how many kids did you kill today?" became a rallying cry that connected personal transformation with political action in ways that British or other European countercultures didn't experience as intensely.

For many young Americans, the choice became stark: comply with a system they saw as fundamentally corrupt, or drop out entirely and create alternative ways of living. Communities like the Love Family represented the mass choosing of the latter option—not just individual rebellion but collective exodus from mainstream society. Dropping out of the system. Almost completely.

Today, we may be witnessing the emergence of a similar perfect storm. After decades of endless American wars following 9/11—from Afghanistan and Iraq to Syria, Yemen, and now the U.S.-

funded genocide in Gaza—a new generation is experiencing profound moral revulsion at their government's active pursuit of extreme violence around the world. The visual documentation of atrocities streaming through social media is creating the same visceral impact that television coverage of Vietnam had on the 60s generation, but with even more immediacy and detail. Combined with the climate crisis, economic precarity, and the rapid advancement of pervasive artificial intelligence that threatens to further alienate humans from authentic experience, we may be approaching another moment of mass disillusionment with mainstream society.

George Douvris, a friend of my father's from their days in Greece, an author of two memoirs, *Crossing Karma Zones* and *Crossing More Karma Zones*, and now my neighbor on the Big Island of Hawaii, offers a particularly insightful perspective on this transformation, having lived through both the original counterculture movement and today's resurgence. When I interviewed him and his wife, Stephanie, in 2025 about their experiences, he identified the dual catalysts that sparked the movement with striking clarity: "The Vietnam War along with LSD were the catalysts that brought a lot of young people in the U.S. and in the world together saying, no more wars. And we, our eyes have opened, our consciousness has opened."

He remembers my father from their time together in a small Greek village where George and his family were living at the time, where despite his unconventional appearance—"he had the longest hair of it all"—my father broke through to the locals: "Your dad had the ability to connect through the heart, his eyes, and just his persona, once you got past the appearance… That worked like a charm."

George's journey from anti-war demonstrations in the American South, where he was threatened by the KKK for his civil rights activism, through communes and various spiritual communities of the 1970s, to his current life in Hawaii as my neighbor, embodies the full arc of the counterculture experience. His perspective is particularly valuable because he witnessed firsthand both the movement's transcendent possibilities and its human limitations—the communities that became "Babylon again, but in a new age guise," as he put it, and the way the "empire struck back" through the War on Drugs and the co-option of hippie ideals into consumer culture.

Just as the hippies in the 60s and 70s sought refuge in rural communes and back-to-the-land movements, today's young people are beginning to explore similar escapes—from tiny houses, to homesteading and permaculture, to digital detox and rejection of AI-dependent lifestyles. The specifics are different, but the underlying impulse remains the same: when the dominant culture becomes morally unbearable and existentially threatening, conscious ethical beings seek alternatives.

George added in a later email to me, worth quoting at length:

> Here are a couple of other factors that I would like to share with you that I feel are significant... During the 60s, the frontier was "wide open" and not as restrictive as today. Individuals and groups could migrate to remote areas both in the US as well as globally without much restriction. There was a feeling of freedom to find and/or create the utopian dream as easy as having a bus ticket or passport, with minimal funds needed. Today places like Goa, Telluride, Maui, Kathmandu, Mexico, etc. are more developed and restrict settlements to just people that have money and are not viewed as a threat to the current social order.
>
> Second big difference is the rise of high tech especially with the internet. During the 60s phone calls were only necessary for emergencies. Important and/or relevant information was shared by letters, alternative as well as mainstream news on television, magazines, word of mouth, etc. When people went to a demonstration or concert, the attention span was more focused. Internet can now be seen as a vital tool of information and communication.
>
> Yet as a tool, one's intention is important. Misinformation, lies, social media promotion of "likes" as ego intoxication and the tsunami of instant electronic information and gratification has become an addictive process with mass usage of advertising psychology that can distract an individual from important priorities while pursuing trivialities.
>
> Thus issues like the genocide in Gaza have to share focus with an onslaught of other information which [dilutes] attention. There could be millions of YouTube "likes" about a celebrity's latest hair style while only a handful of people click on links to learn (and even empathize on a deeper level) about the slaughters being perpetuated in our name.

We'll return to these themes in later chapters as we explore how AI and technology may be driving a new wave of dropping out, and how the hippie insights about consciousness, community, and connection to the land may prove essential for navigating our many current crises.

The Back-to-the-Land Movement: My Mother's Practical Magic

One of the most enduring legacies of the hippie movement was what became known as the "back-to-the-land" movement—a mass exodus from cities to rural areas where young people attempted to live more simply, sustainably, and in harmony with natural cycles. This wasn't just romantic escapism but a practical response to the recognition that modern industrial society was fundamentally unsustainable and alienating.

My own childhood was deeply shaped by this movement, though in our case "back to the land" took on a particularly nomadic form for much of my childhood. My mother embodied the practical wisdom of the movement through her mastery of what others might call traditional crafts but which she understood as essential life skills—spinning wool into yarn with a wooden wheel, dyeing the yarn with natural materials foraged from the countryside, knitting sweaters, gloves, and hats that kept us warm through harsh winters, sewing clothes that lasted for years, growing vegetables wherever we settled, and making wine from whatever fruit or flowers were available.

She made buckets full of cherry wine during our time working in an orchard near Yakima, Washington, transforming the abundant fruit into something that would keep and provide entertainment during the cold eastern Washington winter—at least for my parents, if not us kids. She made elberberry wine back in England and I do remember being allowed some sips from that even though I was only seven or eight at the time. A friend in England once told her she'd never met anyone with so many hobbies, but my mother bristled at that. "I thought of them as life skills, not hobbies," she told me. These weren't recreational activities but survival technologies, essential knowledge for maintaining independence from the consumer economy that the counterculture was rejecting.

I have vivid memories of watching her work the spinning wheel, her hands moving with practiced grace as raw fleece (and sometimes Samoyed dog hair obtained from a friend!) transformed into smooth, strong thread. The rhythmic whir of the wheel became one of the background sounds of a period of my childhood, more constant than my father's philosophical musings. She would gather plants during our travels—creating dyes that brought vibrant colors to her wool. The process was almost alchemical: transforming humble plants into liquid color, then binding that color permanently to fiber through techniques passed down through generations.

Wherever we lived, she would establish small gardens, coaxing vegetables and flowers from the soil. These weren't decorative gardens but food production systems, carefully planned to provide nutrition and variety to supplement our bought (and sometimes scavenged) food. She understood which plants would grow quickly in different climates and seasons, maximizing yield from minimal space and resources.

This practical magic took on special significance amid the chaos of living in a succession of vehicles that served as our mobile home. We started on our grand adventures in the U.S. (we came to the U.S. from England in late 1980, except for my father who had already been in the U.S. for a few years, and ended up just staying, overstaying our visas and later becoming naturalized citizens) in a Volkswagen bus—the iconic hippie vehicle—but as our family grew, we graduated to a "step van" (a small delivery truck) and eventually to a converted yellow school bus that became our home for several years. The bus was fitted out quite nicely inside, with sleeping areas, a small kitchen, and storage for my mother's spinning wheel and baskets of wool. But space was always limited, and my bedroom eventually became the large roof rack that my step-father had built above the bus.

That roof rack bedroom was tolerable during warm weather but became a serious test of survival skills during a brutal winter we spent in eastern Washington, parked in an orchard, where temperatures dropped as low as thirty below zero with wind chill included (I remember these numbers vividly since I was always a numbers and measures geek). I would lie in my sleeping bag on that covered roof rack, feeling the floor conduct the cold directly into my bones, listening to the wind howl across the frozen landscape. I was, however, happy as a bug in a rug—I always had a bit of a survivalist in me. One of our cats would snuggle up with me inside my sleeping bag on the cold nights and that helped both of us get through rather comfortably. It was also my mother's hand-knitted wool sweaters, hats, and socks that kept me relatively comfortable during those nights—practical applications of ancient technologies that no store-bought synthetic materials could match. Looking back on that time, I actually liked it even though it was very far from the mainstream society's notion of the good life, particularly for a kid.

Photo: The yellow schoolbus that was our home for many years, in Washington state. This is the bus parked at our property near Port Angeles, long after it had ceased being mobile.

The back-to-the-land movement represented more than just a lifestyle choice; it was a philosophical statement about the relationship between humans and nature, between traditional knowledge and modern technology, between self-reliance and dependence on industrial systems.

This experience of radical self-reliance (interestingly, a call taken up by Burning Man culture, as I discuss in a later chapter), however chaotic and uncomfortable it sometimes was, taught me lessons that no conventional education could provide. I learned that humans could meet their basic needs through knowledge, skill, and cooperation with natural systems rather than a default and unthinking dependence on industrial production. I saw how traditional technologies—spinning wheels, looms, hand tools—could be more durable and repairable than their modern equivalents. Most importantly, I experienced firsthand how the process of creating what you need for survival can be deeply satisfying in ways that purchasing ready-made goods never is. Keep in mind this was eons before the internet, smartphones, or social media. Life was rather different back then. We didn't even have a TV for most of my childhood. Yes, we read a lot of books. And played games. And went to bed pretty early.

The back-to-the-land movement produced mixed results. Many urban refugees to rural areas lacked the practical skills needed for subsistence living and eventually returned to cities. Some communities devolved into poverty, dysfunction, or exploitation. But the movement also seeded thousands of organic farms, sustainable communities, and small businesses based on traditional crafts and technologies. It validated the possibility of living more lightly on the Earth while maintaining—or even enhancing—quality of life. The key was always real human connection. And that has always been the core aspect of human happiness and thriving.

My mother's spinning wheel, which traveled with us through years of nomadic life, represented the best of this movement: the marriage of ancient wisdom with contemporary necessity, the transformation of raw materials into beautiful and functional objects through patient skill, and the deep satisfaction that comes from creating rather than merely consuming. In our converted school bus rolling down back roads across the western states of America, she was keeping alive technologies and knowledge that industrial society had nearly forgotten—life skills that may prove essential as we face new ecological and social crises.

The hippies' back-to-the-land movement anticipated many of today's concerns about sustainability, local production, and resilience in the face of system breakdown. My childhood amid the organized chaos of mobile homesteading taught me that alternative ways of living aren't just possible but can be deeply fulfilling, even when—or especially when—they require us to rediscover skills and knowledge that our culture has largely abandoned.

The Psychedelic Revolution: Chemical Keys to Consciousness

The role of psychedelics in creating the hippie movement cannot be overstated. As Stevens meticulously documents in *Storming Heaven* (a book I enjoyed immensely in my early 20s as I was exploring the history of the 60s and 70s counterculture), LSD and other psychedelics didn't just alter individual consciousness—they catalyzed a complete worldview transformation that rippled through an entire generation.

The story begins in the laboratories of Swiss chemist Albert Hofmann, who first synthesized LSD in 1938 and accidentally discovered its consciousness-altering properties in 1943. But it was the migration of LSD from laboratory to counterculture that created the conditions for mass consciousness change.

Key figures like Aldous Huxley (and his seminal book, *The Doors of Perception*), Timothy Leary, Richard Alpert (who later took on the name Ram Dass), and Ken Kesey served as bridges between the scientific and cultural worlds. Leary's Harvard Psilocybin Project and later LSD research provided scientific credibility to claims about psychedelics' consciousness-expanding potential. Kesey's Acid Tests, documented by Tom Wolfe in *The Electric Kool-Aid Acid Test*, demonstrated how psychedelics could create new forms of collective experience and artistic expression.

What made psychedelics so revolutionary wasn't just their immediate effects but their tendency to produce lasting worldview changes. Many users reported experiencing the interconnectedness of all life, the illusory nature of ego boundaries, and direct encounters with spiritual dimensions of reality. These weren't just interesting experiences—they were revelations that challenged the fundamental assumptions of materialist culture.

As Stevens notes, psychedelics served as "chemical sacraments" that provided experiential validation for insights that had previously been confined to mystical traditions. The drug culture and the spiritual awakening of the 1960s weren't separate phenomena—they were two aspects of the same search for authentic experience and expanded consciousness.

The mass availability of LSD, particularly through underground chemists like Owsley Stanley, meant that millions of young Americans had access to these consciousness-altering experiences.

By some estimates, over 10 million Americans had tried LSD by 1970. This represented an unprecedented experiment in mass consciousness change, with effects that reverberated throughout society and influenced the formation of countless alternative communities.

The chemicals had opened doors of perception that could never fully close, creating a permanent dissatisfaction with conventional reality and a lifelong quest for communities and practices that honored expanded consciousness.

The Sonic Revolution and Shared Awakening

If psychedelics provided the chemical catalyst for consciousness expansion, music was the delivery system that spread the transformation across an entire generation. The hippie movement was fundamentally a sonic revolution—not just in the sounds being created, but in music's newly recognized power to alter consciousness and unite disparate souls in shared transcendent experience. The very idea of "vibrations" that became so central to hippie philosophy was first and most directly experienced through *music,* in traditional and in new forms, where abstract concepts about resonance and frequency became embodied, ecstatic reality. And available in ever-easier forms through the magic of vinyl and then tapes.

I grew up in 1970s England, in a small town in the southernmost county of Cornwall. I remember fondly my first experiences enjoying music on my own terms: playing Simon and Garfunkel and The Beatles on my parents' record player. I listened to the lyrics of Cecilia: "I got up to wash my face, when I come back to bed, someone's taken my place." I did not understand this at the time (I was only five years old) but I still loved the music. And, yes, I still have a fond spot for everything Beatles, and will always love them. They're deep in my soul.

The 1960s cultural transformation began long before I was born, maybe we can start in the folk clubs and coffee houses of Greenwich Village, where Bob Dylan's acoustic prophecies like "Blowin' in the Wind" and "The Times They Are A-Changin'" gave voice to a generation's growing disillusionment with American empire and injustice. His "Masters of War" stripped bare the military-industrial complex with a fury that academic critiques could never match, while "A Hard Rain's A-Gonna Fall" painted apocalyptic visions that felt more truthful than any newspaper. These songs weren't just protest music—they were consciousness-altering incantations that awakened young Americans to the lies they'd been told. When Dylan went electric at the 1965 Newport Folk Festival, it wasn't a betrayal of the folk tradition but an amplification of its power, a recognition that the movement needed sonic force to match its revolutionary aspirations.

The Beatles' evolution from "Love Me Do" to "Tomorrow Never Knows" traced a parallel psychedelic arc that millions followed, their experiments with Indian instruments, tape loops, and studio innovation showing that popular music could be a vehicle for spiritual exploration. The Grateful Dead emerged from Kesey's Acid Tests as musical shamans, their lengthy improvisations attempting to sonically map the psychedelic experience in real-time. But perhaps most importantly, music became truly communal: Woodstock, Monterey Pop, and countless smaller gatherings weren't just concerts but rituals of collective consciousness, where thousands of people would synchronize their neural oscillations through shared rhythm and melody, literally resonating together in ways that validated the hippie intuition that "we are all one," and, yes, "it's all vibrations… man."

Yet the sonic revolution had roots deeper than white folk clubs and psychedelic rock. The hippies' musical awakening was profoundly shaped by Black American musical traditions—the blues that taught suffering could be transformed into transcendence, the jazz that showed consciousness could improvise new realities in real-time, the soul and R&B that insisted the body and spirit were inseparable. Jimi Hendrix, a Black Cherokee who became psychedelia's guitar shaman, embodied this fusion most dramatically, taking the blues his ancestors had created from oppression and launching it into cosmic space. Sly and the Family Stone created a musical vision of integrated utopia that made racial harmony feel not just possible but funkily inevitable. The Monterey Pop Festival's defining moment might have been when Otis Redding, the soul master from Georgia, converted a crowd of flower children through pure emotional resonance, proving that consciousness expansion didn't require chemicals—sometimes all it took was a voice that could crack open hearts.

The infrastructure of musical transmission was equally revolutionary. FM radio stations like San Francisco's KMPX and New

York's WNEW-FM abandoned the three-minute hit formula, playing entire album sides, connecting seemingly disparate musical traditions, and creating sonic journeys that mirrored psychedelic experience. DJs became guides through consciousness, weaving together Indian ragas, John Coltrane's spiritual jazz, electronic experiments, and rock epics into seamless flows that suggested all human expression was part of one vast composition. Record albums transformed from collections of singles into unified artistic statements—the Beatles' *Sgt. Pepper's*, the Beach Boys' *Pet Sounds*, the Who's *Tommy*—that demanded to be experienced as complete journeys.

Perhaps most importantly, making music became democratized. Suddenly everyone was picking up guitars, forming bands, joining drum circles, chanting in groups. You didn't need to be a professional to participate in the sonic revolution; you just needed to add your voice to the collective *ommmmm*. The hippie movement understood something that modern neuroscience would later confirm: group musical participation synchronizes brains, aligns heartbeats, and creates temporary superorganisms of shared consciousness. Every gathering, from Golden Gate Park be-ins to Vermont communes, pulsed with collaborative music-making that transformed abstract ideas about unity into lived somatic experience.

The Pill and Sexual Liberation

The introduction of the birth control pill in 1960 created a sexual revolution that was inextricably linked to the broader cultural transformation of the 1960s. For the first time in human history, women had reliable control over their reproductive choices, fundamentally altering the relationship between sexuality and procreation. And altering the power dynamics between the sexes.

This pharmaceutical innovation had profound psychological and social effects that extended far beyond birth control. The pill liberated sexuality from its traditional confines within marriage and family formation, allowing for exploration of physical pleasure as an end in itself. This aligned perfectly with the hippie emphasis on immediate experience, authenticity, and, too often, a rejection of delayed gratification.

The Love Family, like many communal movements of the era, reflected these changing sexual mores in complex ways. While ostensibly promoting "free love" and rejection of possessive relationships, the reality was often more complicated, with charismatic leaders like Love Israel exploiting these ideals for personal gratification while imposing different standards on followers.

The sexual revolution also challenged traditional gender roles and power structures. Women's liberation from the fear of unwanted pregnancy enabled greater participation in education, careers, and political activism. The women's liberation movement, which emerged alongside the hippie movement, shared its questioning of authority and commitment to personal freedom.

The pill also contributed to changing attitudes about the body and bodily pleasure. The medical establishment's acceptance of contraception for non-medical purposes implicitly acknowledged that sexuality could be separated from reproduction—a recognition that opened space for broader discussions about consciousness, pleasure, and human potential that would be central to hippie philosophy.

Eastern Winds: Ancient Wisdom Meets Modern Seekers

The importation of Eastern spiritual traditions to America provided the hippie movement with philosophical frameworks that could accommodate their experiences with expanded consciousness. This wasn't cultural appropriation in the shallow sense but a genuine seeking for wisdom traditions that addressed questions Western culture had largely ignored.

Alan Watts played a crucial role in this transmission, serving as perhaps the most eloquent interpreter of Eastern wisdom for Western audiences. His books like *The Way of Zen* and *The Book: On the Taboo Against Knowing Who You Are* provided accessible introductions to concepts that would become central to hippie philosophy—the illusory nature of the separate self, the interconnectedness of all phenomena, and the possibility of awakening to our true nature.

Watts' background was itself a bridge between traditions. Born in England, trained in the Anglican Church, but deeply influenced by Eastern philosophy, he embodied the cultural synthesis that the hippie movement was attempting. His ability to translate Zen koans, Taoist paradoxes, and Vedantic insights into language that spoke to contemporary seekers made him a prophet for the counterculture.

The recognition that beneath our apparent separateness lies a fundamental unity—what Watts described as "the universe playing hide-and-seek with itself"—became a core conviction that would shape his entire approach to life and spirituality.

Several factors converged to make Eastern spirituality attractive to 1960s youth. The Immigration Act of 1965 opened American borders to Asian immigrants, including spiritual teachers like Swami Prabhupada (founder of ISKCON/Hare Krishna), Chogyam Trungpa (Tibetan Buddhism), and various Hindu gurus. Their presence provided direct access to authentic teachings rather than just scholarly interpretations.

The Beatles' highly publicized journey to India to study with Maharishi Mahesh Yogi in 1968 brought Eastern spirituality into mainstream awareness. Their embrace of meditation, vegetarianism, and Indian philosophy signaled that Eastern practices weren't just for academic orientalists but could be living traditions relevant to contemporary seekers. The Beatles were a cultural phenomenon in themselves that was hard to overstate—and when the force of their music and personalities embraced Eastern spirituality, albeit mostly in rather shallow ways, it reverberated widely with youth throughout the Western world.

Perhaps most importantly, Eastern traditions offered practical techniques—meditation, yoga, chanting, breathing practices—that could reliably produce altered states of consciousness without drugs. This provided a sustainable path for consciousness exploration that didn't depend on chemical substances or risk legal consequences. And it felt more authentic, less like taking an elevator to the top of the sacred mountain.

The Love Family's spiritual practices reflected this Eastern influence, incorporating meditation, chanting, and communal rituals that drew from various traditions. Love Israel himself claimed connections to ancient wisdom traditions, including of course messianic Christianity, though the authenticity of these claims was often questionable. Nevertheless, the community's emphasis on spiritual practice and consciousness expansion reflected the broader hippie embrace of Eastern spirituality alongside Christian and Jewish traditions.

Civil Rights and the Template for Resistance

The civil rights movement provided both tactical knowledge and moral inspiration for the hippie movement. The successful use of nonviolent resistance to challenge racial segregation demonstrated that seemingly powerless people could create social change through collective action.

Many hippie leaders had direct experience with civil rights organizing. Jerry Rubin, Abbie Hoffman, and other founders of the Youth International Party (Yippies) had participated in civil rights demonstrations and learned organizing techniques that they later applied to anti-war protests and cultural events.

The civil rights movement also established the moral legitimacy of breaking unjust laws. When Martin Luther King Jr. wrote his "Letter from Birmingham Jail" defending civil disobedience, he provided a philosophical framework that young Americans could apply to draft resistance, drug laws, and other forms of legal rebellion.

Perhaps most importantly, the civil rights movement demonstrated that America was capable of fundamental change when enough people demanded it. This created an optimistic spirit that social transformation was possible—that the American dream could be expanded to include not just racial equality but also peace, environmental protection, and alternative ways of living.

Prosperity and the Luxury of Rebellion

The hippie movement emerged during a period of unprecedented American prosperity. The post-World War II economic boom had created perhaps the first generation in history that could afford to "drop out" of conventional society without facing immediate destitution. They had their parents and grandparents, as well as a new social safety net, to fall back on if needed.

This economic security allowed young Americans to prioritize self-actualization over material accumulation. Abraham Maslow's hierarchy of needs, popular in 1960s psychology, suggested that once basic needs were met, people naturally sought higher purposes like creativity, spirituality, and personal growth. The hippie movement can be seen as the mass pursuit of these higher needs by a generation whose material needs were already secure—at least in terms of falling back on parents as necessary.

My parents' ability to experiment with alternative lifestyles reflected in part this prosperity. Born into relatively comfortable middle-class circumstances (but by no means wealthy) in England, my father, educated and articulate, had the luxury of choosing poverty and spiritual seeking over conventional success. My mother was born in northern England to similarly-situated middle-class parents, a life that she described to me as, while comfortable, just "cold and boring." My parents' choice to effectively dropout of mainstream society was made easier because of the economic safety net that affluent societies provided for their wandering children.

The Love Family attracted many members with similar backgrounds—middle-class young adults who could afford to donate their possessions to the community and live communally because they had resources to fall back on if the experiment failed. This economic privilege was both a strength and a weakness of the hippie movement, enabling experimentation while also limiting its appeal to those without such safety nets.

The prosperity also created what sociologist Kenneth Keniston called a "youth culture"—an extended period of exploration between childhood and adult responsibilities. College enrollments soared in the 1960s, creating large populations of young adults with time and intellectual freedom to question social arrangements.

The Medium is the Message: Technology and Cultural Transmission

The hippie movement was also enabled by new forms of media and communication technology. Television brought images of war, civil rights struggles, and cultural change directly into American homes, creating shared experiences across geographic boundaries.

The expansion of FM radio allowed for longer-form programming that could accommodate the extended compositions of psychedelic rock. Underground newspapers like the *Berkeley Barb*, the *Village Voice*, and *Rolling Stone* created networks of communication that connected scattered countercultural communities.

Rock music became the movement's primary form of cultural expression. Artists like Bob Dylan, the Beatles, the Grateful Dead, the Doors, Janis Joplin, Leonard Cohen, Joan Baez, and Jefferson

Airplane didn't just provide entertainment—they served as cultural prophets whose lyrics articulated the movement's values and experiences. The development of high-quality recording technology and affordable audio equipment meant these messages could reach mass audiences with unprecedented fidelity.

The Beatles' transition from pop to psychedelic music paralleled this evolution from conventional thinking to expanded consciousness. Bob Dylan's electric conversion and often cryptic lyrics provided a soundtrack for cultural transformation that spoke to seekers across continents.

The rise of festivals—from Monterey Pop to Woodstock—demonstrated the power of gathering large numbers of like-minded people in celebratory rather than confrontational contexts. These events served as temporary utopian communities that modeled the hippie vision of peace, love, and cooperation. It was all quite intoxicating. I'm sad to have missed this non-stop party that was the 60s and 70s. Alas, I didn't come into being until 1971, late to the party.

Environmental Awakening: Ecology as Spirituality

Rachel Carson's *Silent Spring* (1962) catalyzed environmental awareness just as the hippie movement was forming. The book's documentation of ecological interconnection and the dangers of chemical pollution provided scientific validation for intuitions about the unity of life that many hippies were experiencing through psychedelics and Eastern spirituality.

The environmental movement that emerged in the 1960s wasn't just about pollution control—it represented a fundamental shift toward seeing humans as part of nature rather than separate from it. This ecological consciousness aligned perfectly with hippie values of harmony, interconnection, and living lightly on the Earth.

The back-to-the-land movement, which saw urban hippies relocating to rural communes, reflected this environmental consciousness. Groups like the Whole Earth Catalog promoted appropriate technology and sustainable living practices that would later influence the environmental movement for decades.

The Love Family's rural properties in Washington State embodied this back-to-the-land ethos. Members attempted to live self-sufficiently, growing their own food and reducing their environmental impact. These efforts were often more idealistic than practical, but they represented genuine attempts to live in harmony with natural systems.

Earth Day 1970, which drew as many as 20 million participants around the country, demonstrated how environmental concerns had moved from the counterculture into mainstream consciousness. The event's emphasis on planetary thinking and ecological awareness reflected the global perspective that psychedelics and Eastern spirituality had helped cultivate.

While the hippie movement was international in scope, its character differed significantly between countries. The British counterculture, exemplified by figures like my father, developed along somewhat different lines than its American counterpart. Without the Vietnam War and draft as catalysts, British hippies focused more on cultural and spiritual transformation than political resistance. The Swinging London scene of the mid-1960s seemed to emphasize fashion, music, and lifestyle rebellion rather than the more politically charged anti-war and anti-establishment activism that characterized the American movement.

The hippie movement also emerged from profound intellectual questioning within American universities. The Free Speech Movement at UC Berkeley in 1964 challenged restrictions on political activity and established students as legitimate participants in academic governance. Beat writers like Jack Kerouac, Allen Ginsberg, and William S. Burroughs had already begun questioning American materialism and conformity in the 1950s. Their emphasis on spontaneity, spiritual seeking, vagabonding, and altered consciousness provided literary models for hippie values.

Academic intellectuals like Herbert Marcuse offered sophisticated critiques of consumer capitalism and "one-dimensional thinking." Marcuse's concept of "repressive tolerance" helped explain how seemingly free societies could actually limit genuine freedom through psychological and cultural manipulation.

The rise of interdisciplinary studies programs reflected growing dissatisfaction with academic specialization and departmental boundaries. Students and faculty increasingly sought holistic approaches to knowledge that could address the complex problems facing society.

Personal and Historical Convergence

These various strands converged in the mid-1960s to create a hot crucible of cultural transformation. The Vietnam War provided the immediate and most obvious catalyst for American youth, but the other factors created the conditions that made mass rebellion possible and sustainable.

The Love Family, where I would later spend part of my childhood, represented an attempt to synthesize all these strands into a coherent alternative society. The community combined spiritual seeking with communal economics, consciousness expansion with practical self-sufficiency, rejection of mainstream values with creation of alternative ones.

That communities like the Love Family often failed to live up to their ideals doesn't diminish the validity of the impulses that created them. As I would learn firsthand, the gap between hippie aspirations and hippie realities was often vast. But the aspirations themselves—for expanded consciousness, deeper connection, and more sustainable ways of living—remain highly relevant and important.

The hippie movement wasn't just a rejection of mainstream values—it was an attempt to create comprehensive alternatives. The combination of psychedelic consciousness expansion, Eastern spiritual practices, sexual liberation, environmental awareness, and political activism created a generally coherent worldview that challenged conventional assumptions about human nature and social organization.

Understanding these multiple strands helps explain both the movement's power and its limitations. The hippies were responding to real problems and genuine insights, but they were also products of specific historical circumstances that couldn't be easily replicated or sustained indefinitely.

As Jay Stevens concludes, the psychedelic revolution represented "the collision of two cultures—one based on the written word and logic, the other on the primacy of experience." This collision created the hippie movement and continues to reverberate through contemporary culture in ways both obvious and subtle. The hippies became hippies because multiple forces converged to make alternative consciousness and alternative living both necessary and possible. Their movement represented not just youthful rebellion but a genuine attempt to create a more conscious, connected, and sustainable way of being human. What's not to like about that?

My own journey from England to traveling in a VW bus with my family, to the Love Family, to living in a school bus, to college at the University of Washington in Seattle, to the US Army, to college again, and then law school, and ultimately to developing scientific theories of consciousness reflects in many ways along this unlikely path the ongoing influence of these hippie insights. The recognition that consciousness might be fundamental rather than emergent, that reality might be more interconnected than we typically assume, that vibration and resonance might be key to understanding both physics and awareness—these ideas have roots in the hippie movement's intuitive explorations.

That many of their specific experiments failed doesn't diminish the validity of their basic insights—insights that continue to influence how we think about consciousness, community, and our relationship with the natural world. As I explore the General Resonance Theory and the nature of consciousness throughout this book, I've found myself returning again and again to themes that the hippies articulated in vernacular terms: the fundamental interconnectedness of all phenomena, the importance of direct experience over received authority, the possibility of expanded consciousness. Of resonance as a universal phenomenon. Of Love as foundational to reality.

The hippie movement created the cultural conditions that made it possible to take seriously questions about consciousness that academic science had largely ignored. Without the counterculture's insistence that consciousness mattered, that altered states revealed genuine insights, that the subjective dimension of experience deserved serious attention, the scientific study of consciousness as we know it today might never have emerged.

In this sense, the hippies were pioneers not just of alternative culture but of alternative knowledge—ways of understanding reality that integrated scientific observation with direct experience, objective measurement with subjective insight. Their failures were often failures of integration rather than failures of vision, inability to synthesize their insights into sustainable practices rather than inability to perceive genuine truths about consciousness and reality. After the ecstasy, the laundry, as Jack Cornfield memorably describes these dynamics in his book by that name. But too many hippies forgot about the laundry, opting instead for more ecstasy without the necessary integration.

Understanding why the hippies became hippies helps us understand not just a historical moment but an ongoing human impulse toward expanded consciousness, deeper connection, and more authentic ways of living. These impulses didn't disappear with the end of the 1960s but continue to manifest in new forms, from the mindfulness movement to psychedelic renaissance to environmental activism to the search for conscious technology.

The hippies were right about many things, but they were also wrong about many others—or, at the least, they failed to sufficiently integrate their insights into their behaviors and lifestyles. Understanding both their insights and their limitations provides a foundation for continuing their unfinished project: the creation of a more conscious, connected, and sustainable human culture. In our current era of ecological crisis, social fragmentation, and technological upheaval, we need their vision more than ever—but we also need to learn from their mistakes and build more wisely on their foundations. I delve more into the shadow side of the hippie movement in a later chapter.

Chapter Five

"Tropical Panpsychism" and Oneness with Everything

In writing this book, I've been reminded just how formative my Army years were to much of my thinking later in life, for many reasons: reading widely during these years (I mean, *very* widely), being in Hawaii for the first time, my first time living in a tropical location, and, last but not least, really not wanting to be in the Army at all—wishing I was instead back in college, learning and progressing on my career. This period of my life was truly seminal and important in providing me the time to engage in literally years of self-directed deep study about many, many topics, including philosophy, spirituality, biology, neuroscience, the classics, and others.

During this time, I developed the inklings of what I later came to call, somewhat tongue in cheek, the school of "tropical panpsychism"—a philosophy of consciousness that was as much influenced by the lush, forgiving landscape I found myself in as by the philosophical and spiritual texts I pored over during those years. There's something about the Hawaiian landscape that makes panpsychism not only intellectually appealing, but intuitive and natural.

When I arrived at Schofield Barracks on Oahu in January of 1992, a strapping lad of 20, I was eager to adventure in this new place, to challenge myself physically, and maybe even to find love (which, alas, remained elusive). Paradoxically, it was in this rigid military context, surrounded by the stunning natural beauty of Hawaii, that I found myself slowly drifting back toward ideas I had dismissed in the temporary certainty about big picture questions that we often enjoy in our late teens.

Unlike the forests of the Pacific Northwest or Northern California where I had spent much of my childhood, Hawaii presented nature largely without menace. There were no predators lurking in the undergrowth, no poisonous snakes hidden beneath rocks, almost no venomous spiders to worry about, and no thorny plants to tear at your skin. And of course it's warm all year round in Hawaii, with not a heater or stove to be seen in most homes. The only real dangers were falling off a cliff or drowning in the ocean—which does happen, of course. And the occasional shark bite or death. This pronounced absence of natural threats created a different relationship with the environment, one where you could surrender to nature mostly without fear.

I remember hiking in the mountains above Schofield and Honolulu, finding spots where the forest opened up to reveal spectacular vistas of the island and the Pacific beyond. In these moments, I often had the warm sensation that I could simply lie back on the moss-covered ground, my naked skin touching the moss and other plants, and be absorbed back into it—not in a morbid sense of decay, but as a kind of blissful reunification. *I am become moss.* The boundary between my body and the earth seemed arbitrary, a temporary separation that could easily dissolve.

This wasn't just some romantic nature worship. It was an intuition about the fundamental nature of reality—that consciousness wasn't confined to human minds but was somehow present throughout the natural world. The vibrant life force of the islands seemed to pulse not just through the plants and animals but through the very rocks and soil. I began to wonder if the hippie talk of my upbringing about "everything is connected, man" might have been pointing, however imprecisely, toward something real.

The military, with its emphasis on order and hierarchy, provided an interesting counterpoint to these ruminations. During the day, I would participate in the regimented life of a soldier, shooting M16s and 155-millimeter Howitzers (big artillery guns), doing a lot of "PT" (physical training, which included my one and only marathon, the Honolulu Marathon in 1992, as well as a fair amount of time in the gym getting "ripped"), where everything had its place in a rigid system. But in my free time, I would escape to landscapes that suggested a very different ontology—fluid, interconnected, alive with awareness at every level.

It wasn't until many years later, during a podcast interview with my friend Christine Marie Mason (on her Rosebud Woman podcast in 2024), that we together coined the term "tropical panpsychism" to describe this phenomenon. We wondered if the philosophical concept of panpsychism—the view that consciousness is fundamental and ubiquitous throughout the universe—might be more intuitively accessible in environments like Hawaii, where nature presents itself as mostly benign and welcoming rather than hostile and threatening—something to be conquered. Perhaps those philosophers who developed panpsychist theories in less hospitable climates had to work harder against the grain of their immediate experience, where nature could seem indifferent to human well-being, or even actively malevolent.

This isn't to suggest that philosophical truths vary with geography—only that certain environments might make particular metaphysical views easier to entertain. The tropical setting of Hawaii created a context where the boundaries between self and world, mind and matter, seemed more permeable than they might in environments dominated by survival concerns.

During my Army years I read Alan Watts' suggestions that "for every outside there is an inside. They go together, like black and white." And various other hints of panpsychism. After leaving the military, I found myself increasingly drawn to philosophical works that articulated and defended versions of panpsychism. David Chalmers' groundbreaking 1996 book, *The Conscious Mind*, provided the intellectual framework I needed to take these intuitions seriously. I read this book shortly after it came out, in Los Angeles after I had left the Army and re-started my college career, at Santa Monica College in 1994. Chalmers' formulation of the "hard problem" of consciousness—explaining how and why physical processes in the brain give rise to subjective experience, an idea for which he is still most famous even today—opened the door to considering alternatives to standard materialist or dualist accounts.

Chalmers pointed out that even if we had a complete physical description of the brain and its workings, we would still face an explanatory gap. Nothing in that description would tell us why these physical processes are accompanied by subjective experience —the feeling of what it's like to be conscious. This "hard problem" suggested that consciousness might be more fundamental than our standard scientific accounts allow. This "hard problem" was largely the classical mind/body problem re-branded in a modern context. The problem is to simply explain the relationship between mind and matter. Why does some matter seem to be associated with mental properties, like in humans and other animals? If there is a line of complexity below which mental properties disappear, where is that line, and why? What if there is no line, but simply increasing complexity of consciousness as matter increases in complexity?

As I wrestled with Chalmers' arguments, I found myself returning to those moments of communion with the Hawaiian landscape. The intuition that consciousness wasn't just an emergent property of complex brains but may somehow be woven into the fabric of reality itself began to seem not just poetically appealing but philosophically defensible.

Later, discovering David Skrbina's comprehensive book *Panpsychism in the West* was a revelation. Here was a scholarly account of a philosophical tradition that had been largely marginalized in modern thought but had nevertheless persisted from the pre-Socratics through figures like Spinoza and Leibniz, to William James and Alfred North Whitehead in the modern era. Learning that brilliant minds throughout history had taken seriously the idea that some form of consciousness might be present in all things made my own intuitions seem less outlandish.

What Skrbina made clear was that panpsychism wasn't some fuzzy New Age concept but a serious philosophical position with variations and nuances, which had been explored by various high-powered thinkers for millennia. Some versions attributed full-blown consciousness to all matter; others suggested more limited forms of subjectivity or experience. What united these views was the rejection of the idea that consciousness emerges suddenly and inexplicably at some arbitrary point in the evolution of complex systems. There is a spectrum of views with panpsychism on one side—mind in some form has always existed and does not emerge—and emergentism on the other side—mind only emerges in nature once sufficient biological complexity has been reached, such as in humans and maybe in other mammals too.

The kicker for me was David Ray Griffin's impressive achievement in *Unsnarling the World-Knot*, a resolution to the hard problem based on Whitehead's process philosophy, which is explicitly panpsychist. As Whitehead states: all things are processes and are "drops of experience."

Galen Strawson would later argue that the emergentist notion of utterly non-experiential matter suddenly giving rise to experience seems to involve a kind of magic that is incompatible with a properly naturalistic worldview. If consciousness exists at all—and my own first-person experience makes that undeniable—then it seems more likely that it must have been present in some form all along. As matter complexified and evolved, so mind complexified and evolved.

The more I explored these philosophical arguments, the more they resonated with that sense of oneness I had experienced in Hawaii and in my early acid trips. It wasn't just that I felt connected to the natural world; it was that this connection revealed something fundamental about the nature of reality itself.

Let me be clear: I'm not suggesting that these moments of oceanic feeling (drug-induced or not) constituted proof of panpsychism. Philosophical positions must be evaluated mostly on their logical merits, and accordance with facts, not their emotional or psychedelic appeal. But these experiences did provide a kind of experiential context that made certain philosophical arguments more compelling than they might otherwise have been. And in this case it felt extremely comforting to accept that I am wholly part and parcel of nature, made from the same stuff, and will easily and gently melt back into nature when my consciousness comes to an end. It was a kind of revelation for me with great emotional impact and appeal.

What convinced me intellectually was the realization that the standard materialist account of consciousness faced insurmountable difficulties. If consciousness is just what brains do, then we face the problem of explaining how subjective experience emerges from entirely non-experiencing matter. This is what philosophers call "radical emergence"—the idea that an entirely new kind of thing can appear at some point in the universe's development.

The problem with radical emergence is that we have no other examples of it anywhere in nature. We see plenty of cases of "weak emergence," where complex systems display properties that couldn't have been predicted from their components—like the wetness of water emerging from non-wet H_2O molecules, or the solidity of a table emerging from mostly empty space at the atomic level. But consciousness seems different. It's not just a matter of complexity; it's a whole new kind of being—interiority, subjectivity, feeling—emerging from stuff that supposedly lacks these qualities entirely.

Panpsychism avoids this explanatory magic by suggesting that consciousness doesn't emerge from non-conscious matter; rather, it's been there all along, becoming more complex and integrated as the physical structures that embody it become more complex and integrated. Just as we accept that matter neither appears nor disappears but transforms, perhaps the same is true of consciousness.

This view still faces challenges, particularly the "combination problem"—explaining how simpler forms of consciousness combine to form the unified consciousness we experience. But this seemed to me a more tractable problem than explaining how consciousness emerges from nothing at all. (My more recent work is focused on reasonable resolutions of the combination problem—and this is just what General Resonance Theory offers.)

Moreover, I began to see how this perspective might bridge the ancient divide between science and spirituality. The intuition of oneness with nature that mystics have reported throughout history might be pointing, however imperfectly, toward a truth about the fundamental nature of reality—a truth that our best philosophical reasoning was now recovering.

There are moments when intellectual conviction and experiential wisdom align, and my embrace of panpsychism represented such a convergence. The Hawaiian landscape had offered an embodied experience of what philosophers like Whitehead were articulating conceptually—that the world is not a collection of dead, inert objects but a community of experiencing subjects, of *organism* everywhere, different in degree but not in kind from our own consciousness.

Decades later, reflecting on that time in Hawaii, I can see how that environment created the perfect conditions for this philosophical journey. The absence of natural threats allowed for a surrender to nature that would have been foolhardy elsewhere. The vibrant, pulsing life of the islands suggested consciousness pervading everything. The contrast with military regimentation, and my strong desire to be back in college and a thirst for understanding, highlighted the inadequacy of purely mechanistic models of reality.

As should be apparent, my philosophical explorations during this period weren't entirely sober. My first experiences with LSD, beginning with a significant trip with my sister and her friends in Seattle, undoubtedly influenced my thinking in this time period at the end of my Army career and in the years after. But I'd push back on any suggestion that these insights were merely drug-induced illusions. Rather, these experiences, like my communion with the Hawaiian landscape, opened perceptual doors that allowed me to consider philosophical possibilities I might otherwise have dismissed.

This transformation was not straightforward or planned in any way. It required both intellectual rigor and experiential openness. In the end, what convinced me wasn't just the emotional appeal of feeling one with nature, but the realization that panpsychism, while seemingly outlandish at first glance, was actually the least problematic of the available metaphysical options. As William James put it, "Consciousness, however small, is an illegitimate birth in any philosophy that starts without it, and yet professes to explain all facts by continuous evolution... If evolution is to work smoothly, consciousness in some shape must have been present at the very origin of things."

This wasn't just philosophical speculation; it was a way of being in the world that transformed my relationship with everything around me. If all things participate in consciousness to some degree, then nothing is truly inanimate, nothing is wholly other. The world becomes not a collection of objects but a communion of subjects, each with its own interior dimension, however rudimentary. It's a very *collegial* view of the world. We're all participating together in the constant flex and flux of nature, in all her marvelous forms.

In this light, tropical panpsychism isn't just an academic position but an ethical orientation—one that calls for reverence toward all manifestations of being, from the grandest celestial bodies to the microscopic processes unfolding within each cell of our bodies. It suggests that the appropriate stance toward reality is not domination but dialogue, not exploitation but partnership.

It's a participatory worldview where all entities, at every organizational level, literally get a say in how reality unfolds in each moment.

And perhaps most importantly, it offers a way of understanding ourselves not as anomalous flashes of awareness in an otherwise dead universe, but, instead, as particularly complex expressions of the consciousness that pervades everything. Our human experience, with its rich textures of thought, emotion, and sensation, represents not a break from nature but its fulfillment in this corner of our galaxy—not an exception to the cosmic rule but its most intricate elaboration, right here, right now.

The Hawaiian landscape had provided the perfect classroom for this education—a place where the boundaries between self and world could soften, where nature presented itself not as adversary but as kin, where the pulse of life was so palpable that it seemed to resonate through everything, even the stones and ocean.

And in living this reality, I discovered something unexpected: far from diminishing the wonder of human consciousness, this perspective enhanced it. By seeing our awareness as continuous with the consciousness pervading the cosmos, human experience became not less but more significant—not an isolated phenomenon but a focal point in the universe's ongoing exploration of its own nature.

This is what I mean by tropical panpsychism—not just an abstract metaphysical position but a way of being that reconciles intellectual rigor with experiential wisdom, scientific understanding with spiritual insight. It's a perspective born of both philosophical analysis and embodied communion with the living world, a view that sees consciousness not as a problem to be solved but as the very fabric of reality itself.

And in embracing this view, I found myself finally at home in the universe—no longer an accidental intruder in an indifferent cosmos but a natural expression of the same consciousness that animates everything from quarks to quasars, in a vastly long unbroken chain of exploration and curiosity. In this recognition was a profound peace that has (mostly) stayed with me ever since, a sense of belonging that transcends any particular place or time.

This, perhaps, is the greatest gift of panpsychism—not just an intellectual solution to philosophical puzzles but an existential homecoming, a realization that we have never really been separate from the world we inhabit, that the boundaries we perceive are often fictions rather than absolute divides. In the end, the hard problem of consciousness dissolves not because we've explained consciousness in terms of something else, but because we've recognized it as the primary reality from which everything else is derived.

Whitehead, who developed one of the most sophisticated panpsychist metaphysics, said it well: "The misconception which has haunted philosophic literature throughout the centuries is the notion of 'independent existence.' There is no such mode of existence; every entity is to be understood in terms of the way it is interwoven with the rest of the universe." It's all connected, man.

In the lush landscapes of Hawaii, this interweaving was not just a philosophical abstraction but a tangible reality, a truth that could be felt in the body and intuited in the heart. And in that feeling, in that intuition, was the seed of a worldview that has continued to grow and evolve throughout my life, providing both intellectual satisfaction and spiritual sustenance.

And in that recognition is a kind of freedom—freedom from the anxiety of separation, from the fear of meaninglessness, from the burden of maintaining an illusory independence or superiority from the rest of reality. We are not isolated minds trapped in alien bodies, nor are we biological machines temporarily housing an enigmatic consciousness. We are expressions of the same conscious energy that animates the entire cosmos, temporary focal points in the universe's ongoing exploration of its own infinite varieties of wiggling and giggling. Yes, we are special. Very special. But for different reasons than much of the Western Judeo-Christian worldview has suggested.

Chapter Six

It's All About Vibrations... Man

It was only after I became more diligent in my intellectual and spiritual questing in my mid-30s and encountered Whitehead's process philosophy and, a bit later, Strogatz's science of sync that I began to see my early psychedelic and cannabis experiences in a new light. Perhaps what I had experienced was a momentary, spontaneous synchronization of neural oscillations—a brief achievement of a particularly powerful and coherent resonant state. In this state, the boundaries that normally separate different aspects of my experience dissolved, allowing for a direct perception of what Whitehead might call the "creative advance" of reality—the endless pulsing of actual occasions into and out of existence. Or something like that. I really have no idea what happened, still.

This reinterpretation of my mystical experience wasn't diminishing it by reducing it to neural processes; rather, it was elevating neural processes by recognizing their potential to provide genuine insights into the nature of reality. If reality is fundamentally vibrational and oscillatory, as both Whitehead and Strogatz suggested from their different perspectives, then states of consciousness characterized by highly coherent oscillations might indeed provide valid insights into this nature.

My intellectual journey toward understanding consciousness as fundamentally vibrational began with psychedelics but developed further over the course of my life with philosophy. Around 2009, and as mentioned previously, I encountered David Ray Griffin's remarkable book *Unsnarling the World-Knot: Consciousness, Freedom, and the Mind–Body Problem*. Griffin presented a clear and compelling case for what he called "panexperientialism"—the view that experience or feeling is inherent in all actual entities, not just complex organisms like humans. A more common term for this view is panpsychism, which we just discussed in the last chapter. This school of thought, derived by Griffin from British philosopher Alfred North Whitehead's process philosophy, offered a potential solution to the mind–body problem that neither reduced consciousness to wholly-mindless physical processes nor posited it as a kind of separate substance.

Griffin's book served as an accessible gateway to Whitehead's notoriously difficult *magnum opus*, his 1926 book *Process and Reality*, which I tackled a year or two later. Reading Whitehead directly was a formidable challenge—his prose is dense and technical, filled with neologisms and complex conceptual structures. Yet the effort, combined with a number of Griffin's books that build upon and clarify Whitehead's system, yielded profound insights that resonated deeply with my experiences and developing theoretical framework.

After discovering and beginning to understand the rich but complex Whiteheadian system, for the first time in my life I felt an acceptance and agreement that this system was very much on the right track. I had rejected many systems as inadequate before this realization. I still don't agree with all aspects of Whitehead's system but my own thinking is about 90 percent congruent with Whitehead's system of thought.

Whitehead's fundamental insight was that reality consists not of enduring substances but of *events*. Events are simply happenings, in time. Events, or "actual occasions", arise, persist briefly, and perish, giving way to new occasions. In Whitehead's system the entire universe is a collection of events arising and perishing and giving rise to the next round of events.

This is a radically different worldview than conventional materialism, which posits unchanging particles careening through the container of spacetime as the basic units of reality. In Whitehead's "process-oriented" system, each actual occasion is a process of becoming that integrates influences from past occasions and creates something novel. This process of creative integration, which Whitehead called "concrescence," or becoming concrete, involves both physical and mental aspects—every actual occasion has both a "physical pole" and a "mental pole."

What struck me most powerfully was Whitehead's portrayal of these actual occasions as inherently oscillatory. Each occasion, such as an electron or atom, pulses into existence, achieves its concrescence (integration), and then perishes, giving way to the next "occasion of experience." This endless rhythm—this cosmic and pervasive pulse—occurs at all levels of reality, from the smallest quantum events to the largest cosmic processes. Reality itself, in Whitehead's vision, is fundamentally rhythmic, vibratory, oscillatory. Even basic particles, which seem to persist over time, are in fact a *serial collection of different but related events*. The electron appears to be the same over time because of this continuity, not because it is in fact an unchanging electron.

This vision of reality as fundamentally *process* rather than *substance* aligned perfectly with the field-based ontology of modern physics and with my experiences of consciousness as flowing and dynamic rather than static. Whitehead's insistence that every actual occasion has both physical and mental aspects offered a way of understanding consciousness as intrinsic to nature rather than an anomalous emergence—exactly the insight I'd been circling around since my spiritual experiences.

Important for my developing theory was Whitehead's concept of "prehension"—the way each actual occasion "feels" or takes account of past occasions. Prehensions could be physical (direct causal influences) or conceptual (taking account of possibilities), but in either case, they involved a kind of resonance—one occasion vibrating in response to the vibrations of others. This resonant relationship between occasions provided a metaphysical foundation for understanding consciousness as fundamentally about vibration and resonance.

Whitehead's vision even accommodated the nesting of conscious entities that I was beginning to see as central to understanding consciousness. In his system, actual occasions can form "societies"—populations of occasions with shared characteristics that could themselves participate in higher-level societies. This nested hierarchy of occasions and societies offered a way of understanding how consciousness might exist at multiple scales simultaneously, from the micro-level of cellular processes to the macro-level of unified human experience in a human body.

Perhaps most remarkably, Whitehead had articulated this vision before the discovery of quantum fields or the development of complex systems theory. Quantum theory was in fact developed right around the same time as he developed his own system, but with quantum theory based on new physics techniques and data, whereas Whitehead's system was a "speculative philosophy" built on logic and mathematics. His metaphysics seemed to anticipate major developments in modern physics and neuroscience, providing a philosophical framework that could integrate these scientific insights into a coherent vision of consciousness and its place in nature.

Griffin's contemporary elaboration of Whitehead's philosophy helped me see how this vision could address specific problems in consciousness studies, particularly the "combination problem" of consciousness that was central to my work. Griffin argued that the combination of lower-level experiences into higher-level ones was possible because experiences aren't like objects that need to be smashed together but more like vibrations that can harmonize and create new patterns.

The synchronization of neural oscillations that I was investigating as a potential neural basis of consciousness could be understood as a physical manifestation of the metaphysical principle that Whitehead had identified: "the many become one, and are increased by one."

This phrase from Whitehead's *Process and Reality*—"the many become one, and are increased by one"—became a touchstone for my understanding of consciousness. It captured perfectly the way resonance allows simpler conscious entities to combine into more complex ones without losing their individual identity. The oscillating neurons don't cease to exist when they synchronize; they continue their individual activity while also participating in a larger, unified pattern. The many (individual oscillators) become one (a synchronized system) and are increased by one (the emergent macro-conscious entity that supervenes on the synchronized system). It is the causal connections between the various constituents that constitutes the higher-level system of consciousness, which does not in any way extinguish the lower-level causal systems, and thus does not extinguish those lower-level conscious entities either.

This Whiteheadian understanding allowed me to see consciousness not as an all-or-nothing phenomenon that suddenly emerges at some arbitrary threshold of complexity, but as existing on a spectrum or continuum. Simple physical systems might have correspondingly simple forms of experience, while more complex, highly integrated systems would support richer, more sophisticated consciousness. The difference was one of degree rather than kind—exactly what we would expect if consciousness is a fundamental feature of reality rather than an anomalous emergence.

Strogatz's "Sync": The Science of Oscillation

Where Whitehead and Griffin provided the philosophical foundation for understanding reality as fundamentally oscillatory, Steven Strogatz's 2003 book *Sync: How Order Emerges from Chaos in the Universe, Nature, and Daily Life,* which I discovered around 2015, provided the mathematical and scientific framework that complemented this philosophical vision, and led soon thereafter to the development of General Resonance Theory. It was massively influential in my academic and philosophical path.

Strogatz, a mathematician at Cornell University, assembled an astonishing array of examples of spontaneous synchronization across vastly different scales and systems. From the synchronized flashing of fireflies in Southeast Asian mangrove swamps to the coordinated firing of cardiac pacemaker cells, from the alignment of menstrual cycles among women living in close proximity to the mysterious coordination of pendulum clocks mounted on the same wall, Strogatz documented a universal tendency toward synchrony that transcends the specific details of any particular system.

What struck me most forcefully about Strogatz's work was how well it aligned with Whitehead's metaphysical vision. Where Whitehead had described reality as composed of pulsing actual occasions forming societies through shared patterns, Strogatz was showing how physical oscillators, such as fireflies, planets, and women's menstrual cycles, naturally fall into synchrony under the right conditions. The "actual occasions" of Whitehead's metaphysics could be understood as nature's fundamental oscillators, and their tendency to form "societies" could be understood in terms of the mathematics of coupled oscillators that Strogatz was developing.

As Strogatz put it in the book's opening lines, "At the heart of the universe is a steady, insistent beat: the sound of cycles in sync." I used this quote at the beginning of our "Easy Part..." paper in 2019, which continues to rack up citations and downloads. This could have been Whitehead speaking about the pulsing rhythm of actual occasions! The convergence was remarkable—a twenty-first-century mathematician and an early twentieth-century philosopher, working from completely different starting points, had arrived at strikingly similar visions of reality as fundamentally rhythmic, with a natural tendency toward synchronization.

Strogatz's mathematics showed how systems of coupled oscillators—any collection of things that cycle regularly and interact with each other—tend naturally to fall into sync under a wide range of conditions. This phenomenon doesn't require central control or coordination; it emerges spontaneously from the interactions between the oscillators themselves. This scientific understanding provided a concrete mechanism for Whitehead's more abstract metaphysical principle that "the many become one, and are increased by one."

For example, consider Strogatz's discussion of the Kuramoto model, which mathematically describes the behavior of large populations of coupled oscillators. Under certain conditions, these oscillators can spontaneously transition from a disordered state, where each oscillator follows its own rhythm, to an ordered state where many oscillators lock to a common frequency. This phase transition from disorder to order isn't designed or planned; it emerges naturally from the dynamics of the system.

Similarly, Whitehead's vision involves actual occasions forming societies through shared patterns of prehension—they "feel" each other and fall into common rhythms. The parallels are striking, suggesting that Whitehead's metaphysical intuitions are being validated by contemporary science.

This integration of Whitehead's process philosophy with Strogatz's mathematics of sync gave me a powerful framework for understanding consciousness as a resonance phenomenon. If reality consists of nested hierarchies of oscillators, from quantum fields to neurons to brain regions to whole organisms, and if these oscillators naturally tend toward synchronization when properly coupled, then consciousness could be understood as a particular kind of resonant integration that occurs in certain complex systems. It suggested that the combination of micro-conscious entities into macro-conscious systems occurs naturally through the same principles that govern the synchronization of any coupled oscillators. And it suggested that the boundaries of a conscious entity are determined by the extent of its resonant integration. There are limits and those limits are defined by resonance.

From Philosophy and Science to Neural Oscillations

With this philosophical and scientific framework in place, I began to see connections to cutting-edge neuroscience research on neural oscillations and their role in consciousness. Researchers like Wolf Singer, Pascal Fries, Christof Koch, and György Buzsáki were developing models of neural function that emphasized the importance of synchronized oscillations. These scientists were showing that consciousness seems to correlate with specific patterns of neural synchrony, particularly in the gamma frequency range (30–100 Hz) associated with perceptual binding and unified awareness.

For example, studies using electroencephalography (EEG) and magnetoencephalography (MEG) had shown that conscious perception correlates with increased neural synchrony, particularly in the gamma range. Anesthetics that render people unconscious also disrupt this synchrony, and techniques like transcranial magnetic stimulation (TMS) that can enhance synchronized neural activity can also enhance aspects of conscious awareness.

These findings aligned perfectly with the resonance framework I was developing. If consciousness is fundamentally about resonant integration of information, then consciousness should indeed correlate with increased neural synchrony. What neuroscientists were measuring as "neural correlates of consciousness" might actually be the physical manifestation of the resonant patterns that constitute consciousness itself.

Different frequency bands might correspond to different levels of conscious integration, with gamma synchrony binding local neural populations into coherent perceptual objects, and slower rhythms, theta (4–8 Hz) in particular being a very strong "organizing wave," integrating these various frequencies into broader conscious contexts.

From Vibrations to Integrated Theory

The convergence of these diverse streams of thought—Whitehead's process metaphysics, Strogatz's mathematics of synchronization,

Leary's psychedelic insights, and contemporary neuroscience—was pointing toward a unified understanding of consciousness that transcended traditional disciplinary boundaries. It wasn't just that these different approaches were compatible; they seemed to be describing the same fundamental reality from different angles.

What Whitehead described as actual occasions pulsing into existence and forming societies through shared patterns, Strogatz might describe as coupled oscillators spontaneously synchronizing, while neuroscientists might measure this as neural synchrony correlated with conscious awareness, and psychonauts might experience it as the dissolution of boundaries and expansion of consciousness. The common thread was resonance—the synchronization of vibrating systems to create *new, emergent patterns that are more than the sum of their parts*.

This integration wasn't just intellectually satisfying; it also made practical sense of my experiences. The dissolution of boundaries between self and world that I'd experienced in Waimea Canyon after my "heroic dose" of acid could be understood as a shift in the resonant patterns of my brain, allowing for greater coherence between my neural oscillations and the oscillating fields of my environment. The spiritual awakening on my brother's couch could be seen as a momentary achievement of a particularly powerful resonant state—a state in which my ordinary boundaries of self dissolved and I experienced a direct connection with the field-like nature of reality.

Most importantly, this integrated framework offered a potential solution to the combination problem of consciousness that had been my central focus. If consciousness involves resonant patterns of information integration, then the combination of simpler forms of consciousness into more complex ones could be understood as a process of achieving shared resonance. Just as pendulums mounted on the same wall will gradually synchronize their swinging, neural populations with the right coupling conditions will synchronize their firing, creating a unified field of activity that corresponds to a unified field of experience.

This resonance-based solution to the combination problem had several advantages over competing approaches. Unlike materialist models that struggle to explain how unconscious components could combine to yield consciousness, the resonance model starts with the assumption that all oscillating systems have at least some rudimentary form of experience, which becomes richer and more complex through resonant integration. Unlike dualist models that struggle to explain how a non-physical mind interacts with a physical brain, the resonance model sees mind and matter as two aspects of the same underlying reality—the inside and outside views of resonant processes.

As my thinking developed, I began to formulate what would later become the core principles of General Resonance Theory:

1. All physical entities resonate—they oscillate, vibrate, cycle between states.
2. When resonating entities interact, they can achieve synchronization, falling into shared resonant patterns.
3. These shared resonant patterns allow for more efficient energy and information exchange between the resonating entities.
4. Consciousness is associated with specific patterns of resonant information integration.
5. The combination of micro-conscious entities into macro-conscious systems occurs through shared resonance.
6. The boundaries of a conscious entity are determined by the extent of its resonant integration.
7. Consciousness exists on a spectrum of complexity corresponding to the complexity of resonant integration a system achieves.

These principles weren't just abstract theoretical constructs; they generated specific, testable predictions about the neural basis of consciousness. If consciousness involves resonant integration, then conscious states should correlate with increased neural synchrony. Disrupting this synchrony should disrupt consciousness, while enhancing it should enhance aspects of conscious experience. Anesthetics that render people unconscious also disrupt this synchrony, and techniques like transcranial magnetic stimulation (TMS) that can enhance synchronized neural activity can also enhance aspects of conscious awareness.

But the resonance framework went beyond these established correlations to suggest a causal mechanism—shared resonance doesn't just correlate with consciousness; it's what allows consciousness to integrate across the brain, creating the unified field of experience that characterizes our consciousness. This mechanism could potentially be tested through interventions that directly manipulate neural synchrony, such as transcranial alternating current stimulation (tACS).

The framework also suggested new approaches to understanding altered states of consciousness. Meditative states, for example, might involve shifts in the resonant patterns of neural activity, with different meditation practices producing different characteristic patterns. Psychedelic states might involve disruptions of normal resonance patterns combined with the emergence of novel ones, allowing for information exchange between neural networks that don't normally communicate. These ideas aligned with emerging research on the neural correlates of meditation and psychedelic experience, which indeed showed characteristic changes in oscillatory brain activity.

The integration of these diverse perspectives—from process philosophy to mathematics, from neuroscience to psychedelic exploration—was yielding a vision of consciousness that was both scientifically rigorous and experientially grounded. It suggested that the hippies' intuition about vibrations wasn't just metaphor or mystical fancy but a genuine insight into the nature of consciousness and reality—an insight that science was only now beginning to formalize and verify.

In the following chapters, we'll explore specific aspects of this resonance-based approach to consciousness in more detail, examining the evidence for consciousness as a field phenomenon, the relationship between different scales of resonance, and the implications of this framework for understanding altered states, artificial intelligence, and the future of consciousness and human societies.

Chapter Seven

Let's Be Scientific!

Testing Theories of Consciousness

This chapter is a little more technical than most, so feel free to skip ahead if it bogs down your progress.

I moved to Los Angeles in mid-1994 after my Army years, and after my brief family visit to Seattle, chasing a childhood crush that had been rekindled earlier that year, at a reunion of the Love Family in Washington State. The "big kids" (most in our early to mid-twenties at this point) of the Love Family had met up in La Push, Washington, for some partying and reminiscing, and some downloads about just how bad and abusive the commune experience had been for some of those kids. Yes, that abuse included some sexual abuse, and a more general abuse of power, particularly by the guy at the top of the food chain, Love himself (yes, that was his name).

I found an apartment in Culver City, one of the many neighborhoods of LA. It was there that I first read Pierre Teilhard de Chardin's masterwork, *The Phenomenon of Man*. Written in the 1930s but not published until after his death in 1955 (due to opposition from the Catholic Church), Teilhard's visionary synthesis of science and spirituality struck me as nothing short of revolutionary. Here was a Jesuit priest and paleontologist offering a coherent account of evolution that encompassed not just biological forms but consciousness itself. And the broad arc of history more generally. With an Omega Point attractor at the end of time. I remain unconvinced by this last part of his sytem, but it is intriguing nonetheless.

What captivated me most was Teilhard's concept of complexity-consciousness—the idea that throughout cosmic history, matter has evolved toward increasingly complex forms, with each increase in complexity accompanied by a corresponding increase in consciousness or "interiority." From the rudimentary interiority of subatomic particles to the self-reflective awareness of humans, consciousness wasn't an anomalous emergence but the natural unfolding of a fundamental cosmic property.

Teilhard proposed that evolution was directional, moving toward that Omega Point—a final state of maximum complexity and consciousness toward which all evolution converges naturally. This teleological view contrasted sharply with the mainstream scientific understanding of evolution as an undirected process driven by random mutation and natural selection. It surely seemed like the history of our corner of the universe has indeed been a march toward ever-ascending consciousness.

What struck me as particularly true was Teilhard's concept of the "noosphere"—a global layer of thought and consciousness surrounding the Earth, emerging from the collective mental activity of humanity. He wrote this decades before the internet, yet seemed to anticipate how human minds would become increasingly interconnected, creating a kind of planetary consciousness. In Teilhard's vision, the noosphere represented a new stage in evolution, as significant as the emergence of life itself.

Reading *The Phenomenon of Man* while simultaneously studying physics and evolutionary theory created a kind of stereoscopic vision—two perspectives that, when combined, revealed dimensions invisible to either alone. Teilhard's account of consciousness as a fundamental property of the cosmos aligned remarkably well with field-based approaches to physics, where reality consists not of isolated particles but interconnected fields in resonant interaction.

Most importantly, Teilhard's work suggested that a scientific approach to consciousness didn't have to be reductionist. We could honor the objectivity and empirical rigor of science while recognizing that consciousness is a fundamental rather than derivative aspect of reality. This balanced perspective has guided my thinking ever since—seeking to ground speculative theories of consciousness in testable predictions and empirical evidence while remaining open to the possibility that consciousness extends beyond what our current scientific frameworks can fully capture. What we can or what we choose to measure will inevitably leave out important aspects of the totality.

In this chapter, we'll explore how the resonance-based approach to consciousness can be tested and refined through rigorous scientific investigation. We'll examine the specific predictions this approach makes, the methods for testing these predictions, and the empirical evidence that either supports or challenges the core claims of General Resonance Theory. By staying grounded in empirical reality while exploring the furthest reaches of consciousness, we can develop a science of consciousness that is both rigorous and expansive.

Testing Theoretical Frameworks

Any scientific theory, no matter how elegant or intuitively appealing, must ultimately be evaluated based on its empirical predictions and explanatory power. This is especially true for theories of consciousness, where the risk of slipping into unfalsifiable speculation is particularly high. So how can we test the resonance-based approach to consciousness we've been developing?

The philosopher of science Imre Lakatos proposed that scientific research programs should be evaluated not on whether individual hypotheses are falsified but on whether the program as a whole is "progressive" or "degenerative." A progressive research program generates novel predictions that are subsequently confirmed, while a degenerative one mainly produces post-hoc explanations for phenomena it failed to predict.

Following Lakatos's approach, we have identified the "core conjectures" of General Resonance Theory—its fundamental, non-negotiable claims—and derived testable predictions from this core. GRT's core conjectures are:

1. The Shared Resonance Conjecture: Shared resonance is the mechanism by which micro-conscious entities combine into macro-conscious entities.
2. The Boundary Conjecture: The boundaries of a macro-conscious entity depend on the velocity and frequency of the resonance chains connecting its constituents.
3. The Nested Consciousness Conjecture: Any biological macro-conscious entity will have various levels of subsidiary/nested micro- and macro-conscious entities.

Each of these conjectures generates specific predictions that can be tested through various methods, from neuroimaging to behavioral studies to phenomenological investigations. Let's examine each in turn.

Testing the Shared Resonance Conjecture

The first conjecture proposes that shared resonance—the synchronization of oscillatory patterns—is what allows simpler forms of consciousness to combine into more complex ones. This generates several testable predictions:

- Prediction 1.1: Conscious states should correlate with measurable synchronization of neural oscillations, while unconscious states should show less synchronization.
- Prediction 1.2: Disrupting neural synchrony should disrupt consciousness, while enhancing synchrony should enhance aspects of conscious experience.
- Prediction 1.3: The transition between unconscious and conscious processing of stimuli should involve a phase transition in neural synchrony.
- Prediction 1.4: Similar principles of synchrony-based integration should operate across different scales of organization in conscious systems.

Considerable evidence already supports these predictions. For instance, numerous studies have shown that conscious perception correlates with enhanced neural synchrony. In contrast, unconscious states like deep sleep, general anesthesia, and certain types of seizures typically show disrupted neural synchrony.

Experiments using transcranial magnetic stimulation (TMS) have shown that disrupting neural synchrony can render stimuli invisible to consciousness, while techniques that enhance synchrony, like certain forms of electrical brain stimulation, can improve aspects of conscious perception.

The transition from unconscious to conscious processing often appears to involve a nonlinear phase transition. For example, when a stimulus gradually becomes visible (as in binocular rivalry experiments), there's a sudden shift in neural synchrony patterns rather than a gradual change, suggesting a phase transition in information processing.

Perhaps most intriguingly, similar principles of synchrony-based integration appear to operate across different scales of biological organization. From the coordinated oscillations of proteins within cells to the synchronized firing of neural populations to the coherent rhythms of brain-wide networks, resonance seems to enable the integration of information across multiple levels.

Testing the Boundary Conjecture

The second conjecture proposes that the boundaries of a conscious entity are determined by the velocity and frequency of the resonance chains connecting its constituents. This generates predictions like:

- Prediction 2.1: The spatial extent of a conscious entity should be limited by the distance that its fastest resonance chain can travel within one cycle of its slowest shared resonance frequency.
- Prediction 2.2: Changes in consciousness (such as expanding or contracting awareness) should correlate with changes in the velocity and/or frequency of relevant resonance chains.
- Prediction 2.3: Different states of consciousness should involve different spatial boundaries, corresponding to different patterns of resonant connectivity.

These predictions can be tested through techniques like high-density EEG, MEG, and invasive electrophysiological recordings, which allow researchers to track the propagation of neural oscillations across the brain. Early evidence suggests that information integration in the brain is indeed constrained by the speed and frequency of neural oscillations, with slower oscillations typically integrating information across larger spatial areas while faster oscillations operate more locally.

For instance, gamma oscillations (30–100 Hz) typically synchronize local neural populations within a few centimeters, while theta oscillations (4–8 Hz) can synchronize activity across distant brain regions. This aligns with the prediction that the spatial extent of information integration (and thus potentially conscious content) depends on the velocity and frequency of oscillatory patterns.

Testing the Nested Consciousness Conjecture

The third conjecture proposes that biological macro-conscious entities contain various levels of nested micro- and macro-conscious entities, creating a hierarchical structure of consciousness. This generates predictions such as:

- Prediction 3.1: Neural activity should show evidence of multiple, nested scales of information integration, each potentially corresponding to a different level of consciousness.
- Prediction 3.2: Disrupting integration at one scale should affect consciousness at that scale while potentially leaving other scales intact.
- Prediction 3.3: Different altered states of consciousness should involve different reconfigurations of this nested hierarchy.

Evidence for these predictions comes from studies of the brain's hierarchical organization and the effects of various consciousness-altering interventions. The brain indeed shows multiple, nested scales of information integration, from local circuits to regional networks to brain-wide systems, each operating at characteristic frequencies and spatial scales.

Selective disruption of integration at different scales produces different effects on consciousness. For example, certain anesthetics primarily affect thalamocortical integration, resulting in unconsciousness, while psychedelics may preserve basic awareness while radically altering the content and structure of consciousness by disrupting certain higher-level integrative processes.

Different altered states—from meditation to psychedelics to dream states—appear to involve different reconfigurations of the brain's nested hierarchy of information processing. For instance, psychedelics often increase entropy and connectivity within and between brain networks, potentially allowing lower-level processes that are usually filtered out to enter awareness.

Behavioral and Phenomenological Testing

Beyond neural measures, the resonance approach to consciousness can be tested through behavioral studies and phenomenological investigations. These methods are crucial because consciousness is fundamentally about subjective experience, not just neural activity.

Behavioral studies might examine how performance on tasks requiring conscious awareness (like reporting the contents of perception) correlates with measures of neural synchrony. If the resonance approach is correct, performance should improve when synchrony is enhanced and deteriorate when it's disrupted.

Phenomenological investigations involve systematically studying first-person reports of conscious experience, often in conjunction with neural measures. This approach, sometimes called neurophenomenology, can help identify correlations between specific aspects of subjective experience and specific patterns of neural synchrony.

For example, studies of experienced meditators have found correlations between reported states of consciousness and measurable patterns of neural oscillation. Different meditation practices (focused attention, open awareness, etc.) produce different patterns of synchrony, and these patterns correlate with distinct phenomenological reports.

Similarly, studies of psychedelic experiences have found correlations between reported aspects of consciousness (like ego dissolution or visual imagery) and specific changes in neural synchrony patterns. The intensity of these experiences often correlates with the degree of change in synchrony metrics, suggesting a direct relationship between resonance patterns and conscious content.

Cross-Species and Developmental Testing

Another approach to testing the resonance theory involves comparing consciousness across species and developmental stages. If consciousness depends on resonant integration, then the capacity for consciousness should correlate with the capacity for neural synchrony across both phylogenetic and ontogenetic development.

Evidence suggests that the capacity for neural synchrony indeed increases with evolutionary complexity, with mammals showing more sophisticated patterns of synchronization than reptiles, and primates showing more complex patterns than other mammals. This aligns with the intuition that consciousness becomes richer and more complex as neural systems evolve greater capacities for resonant integration.

Similarly, the capacity for neural synchrony develops gradually in humans, with full adult patterns of synchronization not emerging until adolescence. This parallels the gradual development of higher cognitive functions and potentially richer forms of conscious experience.

Testing with Artificial Systems

A particularly fascinating avenue for testing the resonance approach involves artificial systems. If consciousness depends on resonant integration rather than specific biological substrates, then artificial systems capable of supporting appropriate resonance patterns might exhibit forms of consciousness.

This doesn't mean that current AI systems like large language models are conscious—they lack the necessary architecture for resonant integration of information. But it does suggest that future systems designed specifically to support field-like, resonant information processing might develop forms of consciousness analogous to biological consciousness.

Such systems could serve as test beds for theories of consciousness, allowing researchers to manipulate parameters of resonance and observe the effects on system behavior. This approach might help resolve long-standing questions about the relationship between information processing and conscious experience.

One of the most controversial aspects of General Resonance Theory is its commitment to a form of panpsychism—the view that consciousness is fundamental and ubiquitous, present in some form at every level of physical reality. How can we test this aspect of the theory?

Direct testing is challenging because we lack reliable methods for detecting consciousness in systems vastly different from ourselves. However, indirect evidence can come from examining whether simpler systems show precursors of the resonance patterns associated with consciousness in more complex systems.

For instance, quantum systems show coherence and resonance phenomena that resemble, at a much simpler level, the resonance patterns observed in neural systems. Molecular systems like microtubules exhibit complex oscillatory behaviors that some theorists have linked to rudimentary forms of information processing. Single-celled organisms show surprisingly sophisticated responses to their environments, including learning and memory.

These observations don't prove that such systems are conscious in any rich sense, but they suggest a continuity of resonant information processing across the spectrum of complexity—exactly what we would expect if consciousness exists on a continuum rather than appearing suddenly at some threshold.

The Challenges of Testing Consciousness Theories

Despite these various approaches to testing, theories of consciousness face unique challenges that theories in other scientific domains don't. The "hard problem" of consciousness—explaining how physical processes give rise to subjective experience—creates a fundamental asymmetry in our investigations.

We can measure correlations between neural activity and reported conscious experiences, but these correlations don't explain why such experiences exist at all. As philosopher Thomas Nagel famously argued, no amount of objective data about bat neurophysiology can tell us what it feels like to navigate the world through echolocation.

This limitation doesn't make consciousness unscientific, but it does mean that the science of consciousness requires a unique methodology—one that integrates third-person measures (like neuroimaging) with first-person reports (like phenomenological descriptions) and theoretical frameworks that bridge the explanatory gap.

The resonance approach attempts to address this challenge by identifying a principle—shared resonance—that operates at both physical and experiential levels. Physically, resonance enables the integration of information across neural systems. Experientially, it corresponds to the unity and coherence of conscious awareness. By focusing on this bridge principle, we might develop a more complete science of consciousness that honors both its physical basis and its irreducible subjectivity.

The Scientific Status of General Resonance Theory

Given the evidence and testing methods outlined above, what can we say about the scientific status of General Resonance Theory? Is it a progressive research program in Lakatos's sense, generating novel predictions that are subsequently confirmed?

The core conjectures of GRT do generate testable predictions, many of which align with existing evidence. And we are finalizing experiments testing some of these conjectures (a couple of papers are in progress, after a few years of gathering data). The emphasis on neural synchrony as crucial for consciousness has been supported by numerous studies across different methodologies. The relationship between resonance patterns and conscious boundaries finds support in research on how different frequencies of neural oscillation integrate information across different spatial scales. The nested hierarchy of consciousness aligns with findings about the brain's hierarchical organization and the differential effects of various consciousness-altering interventions.

However, GRT remains a young theory, and we are just beginning the scientific testing process. These things take time, particularly for new approaches like ours that have historically always had a hard time getting funding, until they're more established, which itself creates a Catch-22. The specific mathematical relationships GRT proposes between resonance parameters and conscious boundaries need more rigorous validation. The extension of consciousness principles across different scales of organization—from subcellular to societal—requires more comprehensive investigation.

Nevertheless, GRT should be considered a progressive rather than degenerative research program. It has generated novel predictions that early research has supported, and it continues to evolve in response to new evidence, rather than resorting to ad hoc hypotheses to protect its core claims.

I was the lead guest editor of a "research topic" (basically an online special issue) of the open-access journal *Frontiers in Human Neuroscience*, completed in 2024, on "electromagnetic field theories of consciousness," of which GRT is an example. This research topic was highly successful in consolidating much of the contemporary science around this class of theories.

We are, at the time of this writing, embarking on a new research topic called Mapping the Resonome, which will flesh out key aspects of GRT's "resonome" concept. The resonome refers to the electrical patterns produced by the brain and body, which in our approach are the actual patterns of consciousness. We look forward to that work, which will probably take two to three years to be completed, and the just-now-unfolding debate about whether consciousness and cognition are based on neural firing or the more gestalt system of nested resonating fields, as suggested by GRT and other electromagnetic field theories of consciousness.

Grounding Theories of Consciousness: A Bayesian Approach

Given the inherent difficulties in testing theories of consciousness, a Bayesian approach might be particularly valuable, and we've suggested this in our published work, in particular in a 2022 paper in *Trends in Cognitive Sciences,* where we propose a "measurable correlates of consciousness" framework in order to create a kind of "consciousness-ometer." Rather than seeking definitive proof or disproof, we can assess how different lines of evidence shift the probabilities of different theoretical frameworks being correct.

In this approach, we start with prior probabilities for different theories based on their intrinsic plausibility and compatibility with existing knowledge. We then update these probabilities as new evidence emerges, increasing the probability of theories that successfully predict the evidence and decreasing the probability of those that don't.

For theories of consciousness, the relevant evidence includes neural correlates of consciousness, behavioral indicators, phenomenological reports, comparative and developmental data, and philosophical considerations about the nature of consciousness and its place in the physical world. More formally, we label these in our paper the "neural correlates of consciousness" (NCC), "behavioral correlates of consciousness" (BCC), and the "creative correlates of consciousness" or CCC.

The resonance approach starts with a moderate prior probability based on its alignment with established principles in physics (resonance as a fundamental physical phenomenon) and neuroscience (the widely acknowledged role of neural synchrony and cross-frequency coupling). As evidence has accumulated showing correlations between conscious states and specific patterns of neural synchrony (what I call the "resonome"), the probability of this approach being on the right track has increased.

However, other theories of consciousness also have significant evidence supporting them, of course, and no single theory currently dominates the field. A Bayesian approach acknowledges this uncertainty while providing a framework for integrating diverse lines of evidence and updating our beliefs as research progresses.

Rather than viewing different theories of consciousness as competing alternatives, however, we might see them as at least in some ways complementary perspectives emphasizing different aspects of the same phenomenon. The resonance approach isn't entirely incompatible with information-based theories like IIT, computational approaches like Global Workspace Theory, or biological theories emphasizing specific neural mechanisms. There is much overlap.

For instance, the resonance approach focuses on how consciousness integrates information through synchronized oscillations, while IIT focuses on the intrinsic properties of systems that integrate information regardless of their substrate. Similarly, the Global Workspace emphasis on global availability of information aligns with the resonance view that consciousness involves large-scale integration of information through synchronized neural activity.

By integrating these perspectives rather than treating them as mutually exclusive alternatives, we might develop a more comprehensive science of consciousness that captures different aspects of this multifaceted phenomenon. All that being said, and while there is overlap between these approaches, GRT captures certain aspects of the physical and biological reality that I feel are not being captured in IIT or GNW.

Staying Grounded While Floating (to Mix Metaphors)

Teilhard de Chardin's work reminds us that scientific rigor and spiritual insight aren't inherently opposed. Indeed, the deepest understanding of consciousness might emerge from their integration. This requires a delicate balance—staying grounded in empirical evidence while remaining open to dimensions of consciousness that transcend our current scientific frameworks.

The resonance approach to consciousness attempts to strike this balance. It's grounded in well-established principles of physics and neuroscience—the ubiquity of oscillatory processes, the importance of synchrony for information integration, the field-like nature of electromagnetic phenomena. Yet it's open to the possibility that consciousness extends beyond what our current scientific methods can directly measure—that it exists across a continuum of complexity and might operate at scales both smaller and larger than individual human brains.

This balanced approach aligns with Teilhard's vision of evolution as the progressive development of complexity-consciousness—a process that includes but transcends purely material mechanisms—if we define "purely material" as that which excludes consciousness. By recognizing consciousness as a fundamental rather than an emergent property, we honor both its physical basis in resonant information processing and its irreducible subjectivity. It's always there and it complexifies as matter complexifies.

As we continue to develop and test theories of consciousness, this integration of scientific precision with philosophical openness offers our best hope for addressing what may be the greatest scientific challenge of our time—understanding the nature of consciousness and its place in the physical world.

Chapter Eight

I Controlled a Huge Freakin' Laser with My Mind!

The Nevada desert at night is a canvas of stars punctuated by fire. During Burning Man, that canvas becomes further adorned with thousands of blinking lights, art cars shaped like dragons and massive pirate ships, pulsing dance domes, and the frequent burst of blue propane flame shooting skyward with a boom and a flash of heat on a cold desert night. In 2010, I stood on this alien landscape with my colleague Jonathan Schooler, a professor at UC Santa Barbara where I was a visiting scholar, and our mutual friend Cassandra Vieten, then the executive director of research at the Institute of Noetic Sciences (IONS).

We were standing in line for a shower, a spanking, or free food, I forget which, plotting what was perhaps the strangest experiment I'd ever been involved with. The basic premise was deceptively simple: see if the collective consciousness of 50,000 Burners could influence the output of random number generators (RNGs) during key moments of mass attention, like the burning of the Man effigy that gives the festival its name, which happens on Saturday night—weather permitting.

"What if we add a laser?" I suggested, the idea materializing suddenly in the dust-filled air between us. "We could connect the output of the RNG to a laser, so people could actually see any deviations from randomness."

Cassandra's eyes lit up. "A visual representation! That would create a feedback loop."

Jonathan, ever the methodical scientist, stroked his dusty chin. "It might actually amplify the effect if people can see it. The collective attention would have something to focus on."

And so, fueled by the creative atmosphere (and let's be honest: plenty of alcohol and drugs) that permeates Black Rock City, our experiment took shape: random number generators monitoring key events, connected to what we jokingly referred to as a "huge freakin' laser" that would visually represent any deviations from randomness.

What had brought three (mostly) respectable academics to this point? For me, it was a long journey from philosophical speculation about consciousness to empirical investigation of its potential effects on the physical world. My interest in panpsychism—the view that consciousness may be inherent in all matter to varying degrees—had led me to wonder about the boundary between mind and matter. If consciousness is fundamental rather than emergent, might it also have effects beyond the confines of individual brains?

Photo: The author at Burning Man with Prof. Jonathan Schooler in 2016.

Dean Radin and the Science of Psi

This is where the work of Dean Radin entered the picture. Radin, the Chief Scientist at IONS, had spent decades conducting experiments on what many would call "psi phenomena"—effects that suggest consciousness might directly influence physical systems. His 2006 book *Entangled Minds*, which I read around 2010, had been transformative for me, not because it convinced me immediately of the reality of psi, but because it showed how such phenomena could be studied scientifically.

Radin's book detailed hundreds of experiments examining various aspects of psi, from telepathy to precognition to psychokinesis. While I remained skeptical about many of these phenomena, one category of research stood out for its methodological rigor and statistical significance: studies showing correlations between human intention and the output of electronic random number generators (RNGs).

Radin didn't start out as a parapsychologist. With degrees in electrical engineering and a background in telecommunications R&D at AT&T Bell Laboratories and GTE Laboratories, his approach to studying psi has always been grounded in rigorous methodology and statistical analysis. It was precisely this background that made his work compelling to me—he brought the tools of hard science to questions that are often dismissed out of hand.

In *Entangled Minds*, Radin presents a comprehensive overview of the experimental evidence for various psi phenomena. The book is notable for its careful attention to methodological critiques and its transparent discussion of effect sizes. Rather than making extraordinary claims based on anecdotal evidence, Radin focuses on the accumulated weight of data from controlled laboratory experiments.

Particularly compelling were the meta-analyses of RNG studies, which combined results from multiple independent experiments to assess overall patterns. One meta-analysis of 515 RNG studies conducted between 1959 and 2000 showed a small but highly significant overall effect, with odds against chance of 10^{35} to 1. This doesn't mean that people can exert great control on random systems with their minds—the effect sizes are typically quite small, around 0.1% deviation from expected randomness. But the statistical consistency across hundreds of studies suggested something consistent and interesting, albeit modest, was happening.

The Slippery Nature of Psi Research

Why hasn't this research had more impact on mainstream science? The reasons are complex and reveal much about both the nature of psi phenomena and the sociology of science itself.

First, there's the issue of replicability. While meta-analyses show consistent effects across many studies, individual experiments don't always replicate. Sometimes an effect appears strongly in one laboratory but weakly or not at all in another. This "decline effect," where positive results become harder to obtain over time, has plagued psi research and fueled skepticism. My colleague, Jonathan Schooler, has spent years studying this phenomenon and he has yet to find any reliable evidence of psi phenomena in his work. He remains skeptical of Radin's conclusion—he likes to "entertain without endorsing" these kinds of things.

Several factors might explain this pattern of decline effects in psi experiments. One is the potential sensitivity of psi effects to psychological and environmental variables that aren't always controlled for or even recognized. Radin and others have noted that experimenter expectation, participant belief, and emotional states may influence outcomes. If psi effects involve some form of consciousness-matter interaction, then the consciousness of everyone involved—including researchers—might matter.

Another challenge is what we might call the "shyness" of psi effects. Unlike most physical phenomena, which behave consistently regardless of whether anyone believes in them or is trying to observe them, psi effects seem to be sensitive to observation and intention in ways that make them methodologically slippery. They tend to appear when not directly pursued and vanish when subjected to rigid scrutiny—a pattern that has led some researchers to propose that psi may involve some inherent element of uncertainty or indeterminacy.

There's also the issue of theoretical framework. Until recently, there was no widely accepted scientific framework that could accommodate psi phenomena without violating established physical principles. This created a high barrier to acceptance—extraordinary claims require extraordinary evidence, as Carl Sagan famously noted, and the threshold for "extraordinary" is much higher when there's no theoretical foundation for the claimed effect.

Another interesting possibility, one which both Schooler and I are somewhat sympathetic to, is that the universe is in fact a simulation and the details of physical laws are not completely built out *a priori*—they are instead developed on an "as-needed" basis as humans and other conscious observers push ever deeper into the fabric of our reality. The simulation may have rough rules that are refined over time *as we experiment* deeper and deeper. More on this set of ideas in later chapters.

Finally, there's the sociology of science itself. Research into psi faces significant institutional and career barriers. Funding is scarce, publication can be difficult, and researchers risk damaging their credibility by association with a field that many consider pseudo-scientific. This creates a self-reinforcing cycle where limited resources lead to methodological limitations, which fuel skepticism, which further limits resources.

Despite these challenges, the body of evidence continues to grow. Particularly promising are studies that combine rigorous methodology with insights from fields like quantum physics, neuroscience, and complexity theory to develop more sophisticated models of how psi might operate.

Our Burning Man Experiments

Our first experiment at Burning Man in 2012 was a first exploration of field consciousness effects. We placed a single RNG near the Man and monitored its output during key events. The results were promising—we found statistically significant deviations from randomness during the burning of the Man, with a p-value of 0.004. This meant the odds of our results occurring entirely due to chance were four in one thousand. This is well beyond the normal p-value of 0.05 in most areas of science. But psi experiments demand a far more stringent standard because of the in-built skepticism among scientists.

Unfortunately, our "huge freakin' laser" had been more aspirational than actual in that first year—technical difficulties had prevented it from functioning as intended. But the data were promising enough to justify a more comprehensive test the following year.

For Burning Man 2013, we ("we" being primarily the IONS team, with me playing a minor supporting role) returned with a much more robust experimental design. This time we deployed six different RNGs using various technologies to generate randomness: some based on quantum effects like electron tunneling in Zener diodes or photon behavior at half-silvered mirrors, others on the timing between emissions of alpha particles from a radioactive source. The diversity of technologies was crucial—if all six systems showed similar deviations at the same time, it would be much harder to dismiss the results as mere technical artifacts.

The data from this second experiment were also showing significant field consciousness effects. Our analysis showed that all six RNGs exhibited similar deviations from randomness at the same time—the moment when the Man's arms were raised, signaling the beginning of the burning ceremony. At this moment every year the crowd goes crazy—after often hours of waiting for the show to begin, circling up in ostentatious "art cars" and standing in the dust in full Burning Man regalia. It's a big moment. The effect we saw in the data was highly unlikely to occur by chance. The probability of this pattern emerging randomly was calculated at less than one in a million.

We went back in 2014 with over 60 RNGs. These data suggested patterns similar to those observed in the previous years, showing real effects from the focused attention of those 60,000 or so people watching the Burn. Unfortunately, Dean decided not to publish these results, so we haven't received much feedback from the scientific community on our data or methods. He told me, when I asked him in 2025 about why the paper was never formally published, that "this series of experiments provided some evidence that collective consciousness influences random physical systems," but "as a purely exploratory effort we felt the results were too preliminary to formally publish." I had really hoped the paper would be published but it wasn't my experiment so it wasn't my call to make.

Our experiments at Burning Man were inspired in part by the Global Consciousness Project (GCP), a larger-scale, long-term research effort initiated by Roger Nelson at Princeton University in 1998. The GCP maintains a network of random number generators at dozens of sites around the world, continuously collecting data to look for correlations between global events and deviations from randomness.

The project has examined numerous events of global significance, from the September 11 attacks to natural disasters to celebrated holidays. The results suggest that during periods of widespread emotional engagement or collective attention, the network of RNGs tends to show more coherence (less randomness) than would be expected by chance.

Critics have raised various methodological concerns about the GCP, including questions about the selection of events to analyze and the specific statistical approaches used. But the project represents an ambitious attempt to study potential mind-matter interactions on a global scale, and its accumulated data provide a valuable resource for ongoing research.

Panpsychism as a Theoretical Framework for Psi

These experimental results, while intriguing, remain challenging to interpret within conventional scientific frameworks. This is where panpsychism and related philosophical perspectives might offer valuable insights.

If consciousness is indeed fundamental rather than emergent—if it's an intrinsic aspect of matter, and based on resonant field effects that literally comprise all of reality, rather than something that appears only in complex brains—then the possibility of direct mind-matter interactions becomes more conceptually plausible. Rather than requiring that consciousness somehow reach "outside" the physical world to influence matter, we might instead be detecting sophisticated interactions within an intrinsically conscious physical reality.

Moreover, if consciousness itself has field-like properties, as suggested by electromagnetic field theories of consciousness proposed by theorists like Johnjoe McFadden, Susan Pockett, me (GRT is a type of EM field theory), and others, then interactions between conscious fields might occur through principles similar to those that govern other field interactions in physics. In other words, "action at a distance" becomes not only possible but inevitable.

The key concept again is resonance—the principle that systems oscillating at similar frequencies can synchronize and amplify each other's effects. Just as a properly tuned radio can detect and amplify even very faint electromagnetic signals, perhaps consciousness fields can interact at long distances when properly "tuned" to each other.

This framework may help to explain several observed characteristics of psi phenomena:

- The small effect sizes: If psi involves subtle field interactions rather than dramatic violations of physical law, we would expect effects to be statistical rather than deterministic, and often quite small.
- The role of intention and attention: If consciousness fields interact through resonance, then focused intention and attention would naturally enhance these interactions, consistent with the observed importance of these factors in psi research.
- The emotional component: Many psi effects seem enhanced by emotional engagement. If emotions involve distinct patterns in consciousness fields, this emotional resonance might facilitate stronger field interactions.
- The distance independence: Some psi effects, like telepathy, appear to operate independently of distance. Field theories could potentially account for this if consciousness fields operate according to principles more similar to quantum entanglement than classical electromagnetic interactions.
- The "shyness": The tendency of psi effects to be elusive under rigid scrutiny might reflect the role of intention and expectation in shaping consciousness fields. If skeptical attention creates field patterns that interfere with the resonance needed for psi effects, this could explain their apparent sensitivity to observation.

William James, the pioneering psychologist and philosopher, once remarked that if telepathy exists, it would be "the only natural function of consciousness which connects it with matter at all likely to throw light on the seat of consciousness in the brain or on its essential nature." This insight resonates with the panpsychist perspective—perhaps what we call psi phenomena are not supernatural but entirely natural, revealing aspects of the consciousness–matter relationship that our conventional frameworks have yet to accommodate.

The physicist and Nobel laureate Brian Josephson has suggested that quantum nonlocality might provide a theoretical basis for telepathy and other psi phenomena. Similarly, David Bohm's implicate order approach (see his astounding book, *Wholeness and the Implicate Order*) proposes that reality has deeper, enfolded dimensions from which our conventional spacetime unfolds—a framework that could potentially accommodate psi without violating established physical principles.

While highly speculative, these approaches represent serious attempts to develop theoretical frameworks that could accommodate psi phenomena within extended but not fundamentally broken physical models.

Randomness and the Nature of Causality

At the heart of RNG experiments like ours at Burning Man lies a profound philosophical question about the nature of randomness itself. What do we mean when we say something is "random"? If RNGs are producing random numbers, what does that mean exactly?

In practice, randomness often serves as shorthand for "we don't know the causes" or "we don't know the pattern." When we flip a coin and call the outcome random, we're really saying that the complex physical dynamics determining how the coin lands are too numerous and subtle for us to track and predict. True randomness —events with no cause at all—may not exist in the classical world.

Quantum randomness presents a different case. According to the Copenhagen interpretation of quantum mechanics, some quantum events, like radioactive decay or which path a photon takes through a double-slit experiment, are fundamentally indeterminate—not just unpredictable in practice but uncaused in principle. Einstein famously rejected this view, insisting that "God does not play dice with the universe."

What if Einstein was partially right, but not in the way he intended? What if quantum indeterminacy represents not an absence of causation but a domain where consciousness can exert subtle influences? After all, as physicist John Wheeler suggested with his concept of the "participatory universe," the act of observation seems to play a crucial role in collapsing quantum possibilities into definite realities.

Several of the RNGs we used at Burning Man were based on quantum processes—either electron tunneling or photon behavior at half-silvered mirrors. If consciousness can indeed influence such processes, even very subtly, then what appears as random noise might actually contain signals of mental influence by pushing outputs away from pure randomness. This was, of course, the point of our field consciousness experiments.

Beyond Burning Man: The Wider Implications

Our experiments at Burning Man represent just one small contribution to a much broader scientific exploration of consciousness and its potential effects on physical systems. But they point toward questions with profound implications for our understanding of reality. If consciousness can indeed influence physical systems in measurable ways, however subtly, this suggests a universe that is more participatory and less mechanistic than conventional science has assumed. It hints at a reality where mind and matter are not fundamentally separate domains but complementary aspects of a unified whole—exactly what panpsychism proposes.

This perspective might help bridge the seemingly unbridgeable gap between the subjective world of experience and the objective world of physical measurement. Rather than treating consciousness as an epiphenomenal ghost in the machine or reducing it to neural activity, we might recognize it as a fundamental aspect of reality that interacts with other aspects in complex, subtle ways.

Such a shift in perspective would have implications far beyond the relatively narrow question of whether minds can influence random number generators. It might inform how we approach fields ranging from physics to biology to medicine to psychology, providing a more integrated framework for understanding phenomena that cross the mind-matter divide.

It might also help reconcile the seemingly irreconcilable worldviews of mainstream science and contemplative traditions that have long emphasized the power of consciousness to shape reality, in a new kind of "deep science," to use Ken Wilber's apt phrase. Not by abandoning scientific rigor or embracing supernatural explanations, but by expanding our scientific models to accommodate the full spectrum of natural phenomena—including those that involve consciousness as a causal factor. I have an almost-finished book of essays (my second), that focuses on an approach that attempts to flesh out Wilber's desire for a new kind of deep science. I hope to publish that book soon.

As William James noted over a century ago, "The universe is wider than our views of it." Perhaps the persistent anomalies of psi research are inviting us to widen our views, to consider the possibility that consciousness plays a more fundamental role in physical reality than our current scientific models acknowledge.

So did I really control a laser with my mind at Burning Man? The answer is both simpler and more complex than that question suggests.

The data from our experiments indicated statistically significant correlations between moments of collective attention and deviations from randomness in electronic systems. These correlations were consistent across multiple technologies and appeared when we would expect them based on our hypothesis. The probability of this pattern occurring by chance was vanishingly small.

Did the collective consciousness of thousands of people somehow directly influence quantum processes in our RNGs? Did it create subtle electromagnetic field effects that interacted with sensitive electronic components? Or was there some other explanation entirely, perhaps involving variables we hadn't considered?

The honest answer is that we don't know—yet. The mystery remains, but it's a mystery that we can continue to explore with the tools of science. I am entertaining without endorsing (at this juncture) the notion that we did in fact discover physical field consciousness effects in our experiments at Burning Man. More work needs to be done.

What's clear is that the boundary between mind and matter may be less absolute than conventional science has assumed. If consciousness is indeed fundamental rather than emergent, as panpsychism suggests, then we shouldn't be surprised to find evidence of mind-matter interactions that challenge our conventional frameworks.

These interactions may never be as dramatic or direct as popular conceptions of telekinesis or "mind control" suggest. The effects seem to be mostly subtle, statistical, and often elusive. But they hint at a deeper integration of consciousness with the physical world than most of modern science has been willing to consider. And if there is a reliable signal, even if small, it could in theory be amplified with technology and perhaps before too long could become a more regular part of our lives and our understanding of the physical world.

As we continue to explore this frontier, we would do well to maintain both open-minded curiosity about what might be possible and rigorous skepticism about what we can claim to know—entertaining without endorsing, as Jonathan likes to say, is a good mantra! The history of science is filled with phenomena that seemed impossible until they weren't—from action at a distance to quantum entanglement to the strange behaviors of complex systems that emerge from simple rules.

Perhaps the relationship between consciousness and the physical world will prove to be another such phenomenon—initially dismissed as impossible, gradually accepted as anomalous, and eventually incorporated into a more comprehensive understanding of reality. Or perhaps we'll discover that what appeared to be mind-matter interactions have entirely conventional explanations that we simply hadn't recognized.

And in a cosmos that has repeatedly proven stranger and more wondrous than our theories predicted, maintaining an open mind about the potential reaches of consciousness seems not just reasonable but essential.

So while I can't claim with certainty that I—or we collectively—controlled a laser with our minds, I can say that our experiments at Burning Man have contributed to a growing body of evidence suggesting that consciousness might have effects beyond the boundaries of individual brains.

Chapter Nine

Tuning the Brain

Energy Medicine & Consciousness

A few years ago, I found myself lying on my very comfortable reading couch in my chartreuse "lava office" at my house in Hawaii, engrossed in Douglas Fields' book *Electric Brain*, a comprehensive history of the science behind electrical brain monitoring and stimulation techniques. Fields, a distinguished neuroscientist at the National Institutes of Health, chronicles how these approaches—from electroencephalography (EEG) to transcranial magnetic stimulation (TMS) to neurofeedback—evolved from fringe curiosities to respected clinical and research tools. What struck me most was how long these electromagnetic approaches to understanding and treating the brain had been marginalized by the mainstream medical establishment, despite compelling evidence of their efficacy.

Fields' account resonated with my own emerging understanding of consciousness as fundamentally electromagnetic in nature. That was, of course, the main reason I had picked up a copy of his book. If complex consciousness arises from synchronized oscillations across neural populations, as General Resonance Theory suggests it does, then directly manipulating these oscillations through electromagnetic means might offer more direct approaches to treating neurological and psychiatric conditions than chemical interventions that affect these oscillations only indirectly.

The history Fields recounts is one of persistent researchers working against institutional skepticism. Hans Berger, who developed the first human EEG in 1924, was initially dismissed by the scientific establishment. His early recordings of brain waves were met with accusations of artifice and fraud. Many of his contemporaries simply couldn't believe that the complex activities of the mind could manifest as simple wave patterns detected through the skull. And yet Berger persisted, meticulously documenting the different frequency bands we now know as alpha, beta, and delta rhythms, laying the groundwork for modern neurophysiology.

Similarly, the pioneers of neurofeedback in the 1960s and 70s faced ridicule and marginalization despite promising results. Researchers like Joe Kamiya, who demonstrated that people could learn to control their alpha waves when given real-time feedback, and Barry Sterman, whose work showed that neurofeedback could reduce epileptic seizures in cats and later humans, were often viewed as engaging in pseudoscience rather than legitimate research.

What these early pioneers were discovering, though they may not have fully articulated it this way at the time, provides a perfect illustration of what philosopher Alan Watts playfully described as two fundamental ways of seeing the world. There are those who see it as full of "spikes"—separate, discrete particles bouncing off each other—and those who see it as fundamentally full of "goo"—flowing, continuous fields that connect everything. Traditional neuroscience has been firmly in the spikes camp, viewing consciousness as arising from discrete neural firing events, which we actually call "spikes." But these electromagnetic pioneers were learning to work with the "goo"—the continuous electromagnetic field dynamics that may be more fundamental to consciousness than we've previously imagined.

From Chemical to Energy Medicine

The dominant paradigm in treating brain disorders for the past half-century has been overwhelmingly chemical. Depression is usually treated with selective serotonin reuptake inhibitors (SSRIs) that increase serotonin levels in synapses. Attention deficit hyperactivity disorder (ADHD) is addressed with stimulants that affect dopamine and norepinephrine pathways. Schizophrenia is managed with antipsychotics that block dopamine receptors.

This chemical approach has undeniably produced significant benefits. Millions of people have found relief from debilitating conditions through psychiatric medications. The introduction of chlorpromazine in the 1950s revolutionized the treatment of schizophrenia, allowing many patients to leave long-term institutionalization and return to their communities. SSRIs have helped countless individuals overcome debilitating depression and anxiety. These successes rightfully earned pharmaceutical approaches their central role in psychiatric treatment.

Yet despite these victories, the limitations of the chemical approach have become increasingly apparent. Many patients don't respond to these medications—roughly one-third of depression patients prove resistant to all available drug treatments. Those who do respond often experience side effects ranging from mildly uncomfortable to severely debilitating—sexual dysfunction, weight gain, emotional blunting, or movement disorders. And even when medications work as intended, they often treat symptoms without addressing underlying causes, requiring indefinite use to maintain their benefits.

These limitations have spurred interest in alternative approaches, including electromagnetic interventions that directly target the brain's oscillatory patterns. Unlike chemical approaches that affect neural function *indirectly* by altering neurotransmitter activity, electromagnetic approaches can *directly* modulate the brain's electrical activity—the very medium through which the brain and body communicate.

In November 2024, I found myself standing at the podium of the Neurotherapy Conference in Santa Barbara, looking out at an audience of clinicians, researchers, and practitioners who had gathered to explore the cutting edge of brain-based therapies. I'd been invited by my friends and colleagues, the power couple comprised of Nick Dogris and Tiff Thompson, to present on electromagnetic field theories of consciousness and their implications for healing.

"Alan Watts once playfully described two fundamental ways of seeing the world," I began, clicking to my first slide. "There are those who see it as full of prickles—separate, discrete particles bouncing off each other—and those who see it as fundamentally full of goo—flowing, continuous stuff that connect everything."

I paused, feeling the audience lean in. "Traditional neuroscience has been firmly in the prickles camp, viewing consciousness as arising from discrete neural firing events, very prickly things. But I'm here to make the case that we need to pay more attention to the *goo*." Goo was, of course, my playful way of describing the emerging understanding of brain and body as comprised primarily of electromagnetic fields and their gooey interactions.

As I clicked through slides showing the evolution of my own thinking and the increasing body of evidence supporting our focus on EM fields, I discussed how many of the therapeutic interventions being discussed at the conference worked by modulating the brain's electromagnetic fields rather than targeting specific neural circuits or neurotransmitter systems. From neurofeedback protocols that train specific brainwave patterns to transcranial stimulation techniques that entrain or "knock out" neural oscillations, these approaches seemed to be working far more with the "goo" than nature's "prickles."

Depression science gives us a good case study of how this prickles-and-goo dichotomy can play out in science and the real world. In 2022, the psychiatric world was shaken when researchers at University College London published the most comprehensive review ever conducted on the serotonin theory of depression (with Joanna Moncrieff as lead author). After examining decades of research involving hundreds of thousands of patients, they found something remarkable: there was no convincing evidence that depression is caused by low serotonin levels. In fact, they discovered that people with depression don't have different serotonin activity than those without depression. Even more surprisingly, when serotonin levels did appear low, it was in people taking antidepressants, not in untreated depressed individuals. The genetic studies were equally damning—massive studies found no difference in serotonin-related genes between depressed and non-depressed people. This revelation is profound because, for decades, doctors have been telling patients they have a "chemical imbalance" that needs correcting with medication. The entire theoretical foundation for prescribing antidepressants to millions of people worldwide has essentially collapsed, even though this is not widely known.

This opens the door for understanding consciousness and mood through entirely different frameworks—like the electromagnetic and resonance-based approaches I've been exploring throughout this book, which may offer more direct ways to understand and heal disorders of consciousness.

Perhaps even more troubling than the collapse of the serotonin theory is what researchers discovered when they gained access to unpublished drug company data through Freedom of Information Act requests. It turns out that most antidepressant trials actually fail—only about 43% show any meaningful benefit over sugar pills, while 57% are failures that rarely see the light of day. When all the data are analyzed together, the placebo effect accounts for about 82% of any improvement people experience on antidepressants. The actual difference between taking an antidepressant and a placebo is less than two points on the standard depression scale—below what experts consider clinically meaningful. Making matters worse, the placebo response has been growing stronger over the decades, rising about 7% every ten years, which means the gap between real drugs and fake ones keeps shrinking. Back in 1982, antidepressants showed a 6-point advantage over placebo; today it's down to just 3 points. Some researchers believe even this small difference might be an illusion caused by patients figuring out they're on the real drug because of side effects, which breaks the blind and creates an enhanced placebo response.

This crisis in psychiatric medicine suggests we need entirely new approaches—perhaps working directly with the brain's natural electromagnetic frequencies and resonance patterns offers a more honest and effective path forward than continuing to prescribe medications based on a discredited theory with marginal benefits at best.

What if many brain disorders in generally aren't primarily chemical imbalances but rather "oscillopathies" that arise from disruptions in the brain's natural electrical rhythms? This perspective suggests a different therapeutic approach entirely: directly manipulating the brain's electromagnetic patterns to restore healthy oscillations, rather than indirect and potentially messy chemical interventions.

Oscillopathies: When Rhythms Go Awry

The term "oscillopathy" was coined to describe disorders characterized by abnormal neural oscillations. This framing suggests that many neurological and psychiatric conditions might be understood primarily as disruptions in the brain's natural rhythms rather than chemical imbalances—or in Watts' terms, as problems with the "goo" rather than just the "spikes."

Consider epilepsy, perhaps the clearest example of an oscillopathy. During a seizure, large populations of neurons begin firing in hypersynchronous patterns—excessively synchronized activity that disrupts normal brain function. This aberrant synchrony can be directly observed in EEG recordings, which show characteristic spike-and-wave patterns. While chemical interventions (antiepileptic drugs) can help control these abnormal oscillations, they do so indirectly and often with significant side effects. Direct modulation of the brain's electrical activity, through approaches like neurofeedback or electrical stimulation, targets the oscillatory disturbance more directly.

Parkinson's disease offers another compelling example. The motor symptoms of Parkinson's—tremor, rigidity, and bradykinesia (slowness of movement)—correlate with abnormal beta oscillations (13–30 Hz) in the basal ganglia and motor cortex. These aberrant rhythms effectively "jam the signal" of normal motor control, like static disrupting a radio broadcast. Deep brain stimulation can dramatically reduce these symptoms by delivering high-frequency electrical stimulation to key nodes in the motor circuit, effectively overriding the pathological beta rhythms. The effects can be immediate and profound—a patient's hand tremor may cease within seconds of activating the stimulation.

Schizophrenia presents a more complex picture but still shows characteristic oscillatory abnormalities. Patients with schizophrenia typically show reduced gamma synchrony (30–100 Hz), particularly during cognitive tasks. This reduced high-frequency band synchronization may underlie the cognitive disorganization, perceptual distortions, and reality-testing deficits that characterize the disorder. The inability to generate coherent gamma oscillations may prevent the proper integration of information across different brain regions, resulting in hallucinations, delusions, and thought disorder. While current antipsychotic medications primarily target dopamine receptors, emerging evidence suggests that directly enhancing gamma synchrony might more directly address the core information-processing deficits in schizophrenia.

Depression, too, shows characteristic oscillatory signatures. Depressed patients often exhibit asymmetry in alpha activity (8–12 Hz) between the left and right frontal regions, with relatively greater right-sided activity. This asymmetry correlates with negative emotional states and reduced approach motivation. Neurofeedback protocols targeting this asymmetry have shown promise in alleviating depressive symptoms, as has TMS treatment targeting the left dorsolateral prefrontal cortex to rebalance this activity. These electromagnetic approaches directly address the oscillatory imbalance rather than simply modulating neurotransmitter levels.

ADHD similarly shows distinct oscillatory patterns—typically increased theta activity (4–8 Hz) and decreased beta activity in frontal regions. This oscillatory signature suggests impaired attention regulation and executive function. Neurofeedback protocols that train individuals to decrease theta and increase beta activity have shown efficacy comparable to stimulant medication in some studies, with the advantage of promoting lasting changes rather than requiring ongoing treatment.

Even autism spectrum disorders show atypical patterns of neural synchrony, particularly in gamma frequencies during sensory processing and social cognition. These altered synchronization patterns may contribute to the sensory hypersensitivities, perceptual differences, and social cognitive challenges characteristic of autism. While current treatments primarily focus on behavioral interventions, approaches that directly target these oscillatory abnormalities might offer new therapeutic avenues.

These patterns suggest that many brain disorders might be better understood as specific disruptions in the brain's resonant patterns rather than solely as chemical imbalances. This reframing opens the door to therapeutic approaches that directly target these disrupted oscillations—working with the "goo" rather than trying to fix the "spikes."

From a resonance perspective, this makes intuitive sense. If consciousness involves specific patterns of resonant information integration, then disorders of consciousness—whether severe conditions like coma or more subtle conditions like depression or anxiety—would almost surely involve disruptions in these resonant patterns. Restoring healthy patterns of resonance would then restore healthy conscious function. The key question becomes how we can best restore these healthy resonant patterns. There's a growing body of science and practice around this issue.

The shift from a chemical to an electrical understanding doesn't invalidate the importance of neurotransmitters and other chemical factors. Rather, it suggests a more integrated view where chemical and electromagnetic aspects of brain function are seen as complementary rather than competing explanations. Neurotransmitters and neuromodulators influence the brain's oscillatory patterns, and oscillatory activity affects neurotransmitter release and receptor sensitivity. The two domains interact continuously in healthy brain function and can both go awry in pathological states.

The oscillopathy framework also helps explain why some traditional healing practices might have genuine efficacy despite lacking, in many cases, modern scientific studies. Practices like rhythmic drumming, chanting, breathing, or repetitive movement, found in shamanic and indigenous healing traditions worldwide, seem to modulate brain and body oscillatory patterns, creating beneficial rhythms. Modern practices like mindfulness meditation and breathwork similarly affect neural oscillations, with experienced meditators showing enhanced alpha and gamma synchrony. Even music therapy can be understood as a way of using external rhythmic stimuli to entrain and normalize neural oscillations. These approaches, often dismissed as placebo effects or psychological support, may actually work through direct effects on the brain's electromagnetic patterns. More science is needed on these practices.

Neurofeedback: Learning to Tune Your Own Brain

Among electromagnetic interventions, neurofeedback stands out for its non-invasive nature and its emphasis on self-regulation. In neurofeedback, individuals receive real-time feedback on their brain activity, typically measured via EEG, and learn to modify this activity through a process similar to operant conditioning.

The history of neurofeedback arguably begins in the 1960s with Joe Kamiya's pioneering work at the University of Chicago. Kamiya discovered that people could learn to distinguish between alpha and non-alpha states and eventually control their alpha production when given simple feedback—usually an auditory tone that would sound when they produced the desired brain activity. This discovery challenged the prevailing view that brain activity was entirely unconscious and automatic, suggesting instead that people could learn conscious control over their neural oscillations with appropriate feedback.

In the early 1970s, Barry Sterman at UCLA made another crucial discovery while studying sleep in cats. Cats trained to produce a specific brain rhythm (12–15 Hz over sensorimotor cortex, later called the sensorimotor rhythm or SMR) unexpectedly showed increased resistance to seizures when later exposed to a seizure-inducing chemical. This serendipitous finding led Sterman to develop SMR neurofeedback for epilepsy in humans, showing that people with epilepsy could significantly reduce their seizure frequency by learning to enhance this protective rhythm.

Despite these promising early findings, neurofeedback remained marginal for decades, overshadowed by the rapid growth of psychopharmacology and viewed skeptically by mainstream medicine. *If it didn't involve a pill, it wasn't taken seriously.* Only in recent years has neurofeedback begun to gain wider acceptance, supported by a growing body of research and increasingly sophisticated technology for measuring and providing feedback on brain activity.

Modern neurofeedback typically involves placing electrodes on the scalp to measure electrical activity from the brain. This signal is then processed to extract relevant features—usually power in specific frequency bands at specific locations. For example, a protocol for attention enhancement might measure the ratio of beta (associated with focused attention) to theta (associated with mind-wandering) activity over the frontal lobes. This information is presented to the individual in real time through visual or auditory feedback—perhaps a spaceship on a screen that moves higher when beta increases relative to theta, or music that plays more clearly when the desired brain state is achieved.

Through repeated sessions, individuals can learn to produce the targeted brain activity more consistently and intentionally. Initially, this learning happens through trial and error—people notice what mental strategies seem to produce the desired feedback and gradually refine these approaches. With practice, the ability to produce the targeted brain state becomes more automatic, eventually transferring to daily life situations even without feedback.

What makes neurofeedback particularly interesting from a resonance perspective is that it doesn't just target symptoms but directly addresses the resonant patterns that underlie consciousness itself. As Fields notes in his book, the technology has now been validated through numerous controlled studies for conditions ranging from ADHD to anxiety to PTSD. We're seeing the emergence of what he calls "precision neuromodulation"—the ability to target specific brain rhythms for therapeutic effect.

For attention deficit disorder, theta/beta neurofeedback training helps individuals increase beta activity (associated with focused attention) while decreasing theta activity (associated with mind-wandering). This effectively helps them shift their dominant resonance pattern from one associated with distractibility to one associated with sustained attention.

For anxiety disorders, alpha and alpha–theta protocols help individuals achieve the relaxed-but-alert states characteristic of alpha predominance, counteracting the high-beta patterns associated with anxiety and rumination. From a resonance perspective, this represents a shift from high-frequency, low-amplitude desynchronized patterns to more coherent, synchronized oscillations that integrate information more effectively.

For PTSD and trauma recovery, neurofeedback addressing abnormal resonance patterns in networks like the default mode network (DMN) and salience network has shown remarkable efficacy. These approaches recognize that trauma disrupts normal resonance patterns between brain regions, creating either hyper-coherence (regions stuck in overly rigid synchronization) or hypo-coherence (regions unable to synchronize appropriately).

For insomnia, training slow-wave oscillations can help restore the natural resonance patterns of sleep, helping individuals transition between sleep stages more effectively. This approach recognizes sleep as a complex orchestration of different resonance patterns rather than simply "turning off" consciousness.

Research on neurofeedback shows promising results for various conditions. For ADHD, multiple controlled studies have shown significant reductions in inattention, hyperactivity, and impulsivity after neurofeedback training, with effects comparable to stimulant medication in some cases. Importantly, these benefits persist after training ends, suggesting that neurofeedback promotes lasting changes in neural function rather than requiring ongoing intervention.

For epilepsy, neurofeedback training to enhance SMR rhythm and suppress abnormal brain activity has reduced seizure frequency by 50% or more in many individuals with epilepsy resistant to medication. Some patients have achieved complete seizure freedom, though this outcome is less common.

Anxiety and depression have also shown responsiveness to neurofeedback, with protocols targeting alpha asymmetry (the pattern of relative left/right frontal alpha power associated with approach/withdrawal motivation) or enhancing alpha/theta ratios to promote relaxation and emotional processing. These approaches have shown particular promise for anxiety disorders resistant to conventional treatments and for addressing the emotional dysregulation that underlies many mood disorders.

Sleep disorders represent another area where neurofeedback has shown efficacy. Protocols that enhance slow-wave activity have improved sleep quality, reduced time to fall asleep, and increased total sleep time in individuals with insomnia. For those with sleep apnea, neurofeedback targeting sensorimotor rhythms can reduce respiratory events by improving upper airway muscle tone during sleep.

Even trauma-related conditions have responded to neurofeedback interventions. Alpha–theta neurofeedback, which promotes a deeply relaxed but alert state similar to that achieved in meditation, has reduced PTSD symptoms in military veterans and other trauma survivors. This approach seems to facilitate processing of traumatic memories in a state where emotional reactivity is reduced but awareness remains intact.

What's particularly intriguing about neurofeedback from a resonance perspective is that it directly engages the individual's consciousness in modifying its own substrate. The conscious mind learns to tune its own resonant patterns, creating a feedback loop between the experience of consciousness and its physical basis in neural oscillations. This self-regulatory aspect distinguishes neurofeedback from other electromagnetic interventions. While approaches like TMS or tDCS apply external fields to modulate brain activity, neurofeedback leverages the brain's intrinsic capacity for self-regulation. It doesn't impose patterns from outside but helps the brain recognize and modify its own dysfunctional patterns. This might explain why neurofeedback's effects often persist after training ends—the brain has learned new patterns rather than temporarily responding to external modulation.

Modern neurofeedback has evolved far beyond the simple single-channel approaches of its early days. Advanced systems now offer quantitative EEG assessment (comparing an individual's brain patterns to normative databases), 3D source localization (identifying the neural sources of surface EEG patterns), and real-time functional connectivity training (targeting the synchronization between different brain regions rather than just activity in single regions). Some systems combine neurofeedback with other modalities like heart rate variability training, acknowledging the intimate connection between brain and body rhythms in determining our mental and emotional states.

As technology continues to advance, EEG-based neurofeedback is being joined by other neuroimaging modalities. Real-time functional magnetic resonance imaging (rtfMRI) neurofeedback allows individuals to learn control over activity in deep brain structures not accessible to EEG, while near-infrared spectroscopy (NIRS) offers a middle ground—more precise spatial resolution than EEG but more portable than fMRI. These emerging approaches promise even more targeted and effective neurofeedback interventions in the future.

Despite its promising results and growing sophistication, neurofeedback still faces challenges in gaining mainstream acceptance. Research quality has been variable, with many early studies lacking appropriate control conditions or sufficient sample sizes. The field also suffers from protocol heterogeneity—different practitioners may use substantially different approaches for the same condition, making it difficult to evaluate efficacy across studies. Additionally, the mechanism of action remains incompletely understood, though the growing recognition of oscillopathies as a framework for understanding brain disorders is helping clarify how and why neurofeedback works.

Nevertheless, the accumulated evidence increasingly supports neurofeedback as a legitimate and often effective intervention for various brain-based conditions. As our understanding of the brain's oscillatory dynamics continues to deepen, neurofeedback is likely to play an expanding role in the treatment of neurological and psychiatric disorders—and potentially in optimizing brain function even in healthy individuals.

Neurostimulation: External Tuning of Neural Oscillations

While neurofeedback works by helping individuals learn to regulate their own brain activity, transcranial stimulation techniques and related neurostimulation techniques directly modulate neural oscillations through externally applied electromagnetic fields. These approaches include transcranial magnetic stimulation (TMS), transcranial direct current stimulation (tDCS), and transcranial alternating current stimulation (tACS)—each offering different ways to work with the brain's electromagnetic "goo."

Transcranial magnetic stimulation stands as perhaps the most established of these techniques. Developed in the 1980s, TMS uses rapidly changing magnetic fields to induce electrical currents in neural tissue. A TMS device consists of a wire coil through which a powerful electric current is passed. This current generates a magnetic field that passes through the skull and induces a secondary electric current in the targeted brain region. Depending on the stimulation parameters, this induced current can either excite or inhibit neural activity.

Initially developed as a research tool for mapping brain function (by observing what happens when specific regions are temporarily activated or inhibited), TMS has evolved into an established treatment for clinical conditions. The magnetic field used in TMS passes through the skull painlessly and non-invasively, creating small electrical currents in targeted neural tissue. These currents can either enhance or suppress neural activity depending on the stimulation parameters. For treatment-resistant depression, TMS typically targets the left dorsolateral prefrontal cortex, a region often showing reduced activity in depressed patients. By stimulating this region repeatedly over several weeks, TMS can normalize activity patterns and substantially reduce depressive symptoms—often in patients who've failed to respond to multiple medication trials.

Transcranial direct current stimulation (tDCS) represents another approach, applying weak electrical currents directly to the scalp. While simpler and less precise than TMS, tDCS has shown promise for various conditions, from depression to chronic pain to stroke rehabilitation. The gentle electrical current subtly shifts the resting potential of neurons in the affected regions, making them either more or less likely to fire. This modulation can enhance neuroplasticity, the brain's ability to reorganize and form new neural connections, potentially facilitating learning, recovery from injury, or the breaking of maladaptive patterns.

Perhaps most aligned with the resonance theory of consciousness is transcranial alternating current stimulation (tACS), which applies oscillating electrical currents at specific frequencies to entrain brain waves to those frequencies. If consciousness involves specific patterns of neural synchrony across different frequency bands, tACS offers a direct method for manipulating these patterns. Early research suggests potential applications in enhancing memory, attention, and creative problem-solving, as well as treating conditions like insomnia and Parkinson's disease by normalizing disrupted oscillatory patterns.

Newer forms of transcranial stimulation are emerging that offer even more precise targeting and control. Transcranial pulsed current stimulation (tPCS) uses pulsed rather than continuous current, potentially offering more specific effects with fewer side effects. Temporal interference stimulation applies high-frequency currents through multiple electrode pairs, creating an interference pattern that can stimulate deep brain regions non-invasively.

As these technologies advance, they offer increasingly precise tools for modulating the brain's oscillatory dynamics and potentially addressing disorders of consciousness at their roots. However, they also raise important questions about the ethics of manipulating consciousness through direct electromagnetic intervention. Who should have access to these technologies? For what purposes? How do we balance potential benefits against risks of unintended consequences? These questions become particularly urgent as simpler forms of transcranial stimulation become available as consumer devices, potentially allowing individuals to modulate their own brain activity outside of clinical supervision.

What's particularly striking is how this scientific validation has occurred without researchers necessarily understanding why these interventions work. As Fields documents, many neurofeedback and stimulation protocols were developed empirically—practitioners noticed that certain frequencies produced certain effects and refined their approaches through trial and error. Only later did the theoretical understanding catch up, revealing that these interventions were working by modulating the brain's resonance patterns—tuning the "goo" rather than targeting individual "spikes."

Deep Brain Stimulation: Precision Tuning of Neural Circuits

While neurofeedback and transcranial stimulation work non-invasively from outside the skull, deep brain stimulation (DBS) involves surgically implanting electrodes into specific brain regions to deliver precisely targeted electrical stimulation. This invasive approach offers unparalleled spatial precision in modulating neural activity, though with the obvious drawbacks of requiring brain surgery with its associated risks.

DBS was first developed in the late 1980s as a treatment for movement disorders, particularly Parkinson's disease and essential tremor. Its development built on the earlier discovery that lesioning specific regions in the basal ganglia could dramatically reduce tremor and other movement symptoms. Rather than creating permanent lesions, DBS offered an adjustable, reversible alternative—electrical stimulation that could modulate neural activity without destroying tissue.

For Parkinson's disease, DBS typically targets one of two main regions: the subthalamic nucleus or the globus pallidus interna, key nodes in the basal ganglia circuits that regulate movement. The procedure involves implanting thin wire electrodes precisely into the target structure, guided by detailed brain imaging and electrophysiological mapping to ensure accurate placement. The electrodes connect to a pulse generator (similar to a cardiac pacemaker) implanted under the skin in the chest, which delivers carefully calibrated electrical pulses to the target region.

The effectiveness of DBS for Parkinson's disease can be remarkable. In many patients, turning on the stimulation produces an immediate reduction in tremor and rigidity, allowing them to move more freely and perform tasks that had become impossible. The dramatic nature of this effect is evident in countless videos showing patients whose severe tremors cease almost instantly when their stimulator is activated. This rapid effect suggests that the stimulation directly disrupts pathological oscillations rather than gradually altering neural structure or chemistry.

The mechanism by which DBS exerts these effects remains incompletely understood and likely varies depending on the target region and stimulation parameters. One prevalent theory is that high-frequency stimulation effectively masks or disrupts the pathological oscillations that characterize conditions like Parkinson's disease. In essence, DBS may work by overriding the abnormal beta rhythms with its own imposed pattern, restoring more normal patterns of neural activity.

As DBS has proven effective for movement disorders, researchers have expanded its application to psychiatric conditions. For treatment-resistant depression, DBS targets regions like the subgenual cingulate cortex or the nucleus accumbens, brain areas implicated in mood regulation. Early studies have shown promising results, with some patients experiencing dramatic improvements in mood and functioning after years of failed treatments.

Similarly, DBS has been investigated for obsessive-compulsive disorder (OCD), targeting circuits involving the anterior limb of the internal capsule or the subthalamic nucleus. The repetitive, intrusive thoughts and compulsive behaviors that characterize OCD may reflect abnormal oscillatory patterns in specific neural circuits, which DBS can normalize through targeted stimulation.

Even the peripheral nervous system offers targets for electromagnetic intervention. Vagus nerve stimulation (VNS), achieved by implanting a small device that delivers electrical pulses to the vagus nerve in the neck, has proven effective for epilepsy and treatment-resistant depression. The vagus nerve serves as a major communication pathway between brain and body, and stimulating it can modulate brain activity in regions crucial for mood regulation and seizure control.

One of the newest electromagnetic approaches, photobiomodulation, uses specific wavelengths of light to influence cellular function, including neural activity. Early research suggests potential applications in treating traumatic brain injury, neurodegenerative disorders, and mood disorders by enhancing mitochondrial function, reducing inflammation, and promoting neuroplasticity.

DBS offers one of the most precise tunings of neural oscillations currently available. By targeting specific nodes in neural networks, it can modulate oscillatory patterns with spatial and temporal precision unmatched by other approaches. This precision comes at the cost of invasiveness, limiting its use to conditions severe enough to justify brain surgery. But it provides proof of principle for the therapeutic potential of directly manipulating neural oscillations.

As technology advances, DBS systems are becoming more sophisticated. Newer systems offer directional electrodes that can steer current in specific directions rather than radiating in all directions equally, allowing for more precise spatial targeting while minimizing side effects. Some systems now include sensing capabilities, monitoring the brain's electrical activity and automatically adjusting stimulation parameters in response to changes in neural state—creating a closed-loop system that can adapt to the brain's moment-to-moment needs.

The future may bring less invasive approaches that combine the precision of DBS with the safety of non-invasive methods. Techniques like temporal interference stimulation show early promise for targeting deep brain structures without surgery by using interfering electrical fields applied from outside the skull. If successful, such approaches could make precise modulation of deep brain circuits accessible to a much wider range of patients with less risk.

Pulsed EM Fields Therapy

Pulsed EM Fields (PEMF) therapy uses magnetic pulses to help the body heal itself. Think of it like a tuning fork for your cells—the electromagnetic waves pass through your body and help restore normal cellular function. The FDA first approved this technology in 1979 for healing broken bones that wouldn't mend on their own, and since then, researchers have discovered it can help with many other conditions. A 2024 study showed PEMF significantly reduced shoulder pain and improved movement for months after treatment stopped. What's especially interesting is how it seems to work differently for different conditions. People with fibromyalgia—a condition involving widespread pain and brain fog—responded better to PEMF applied to the head than people with regular joint or muscle pain. This suggests the therapy might work by helping restore normal brain wave patterns that have gotten out of sync.

One of the most intriguing aspects of PEMF is that there's no single "dose" that works for everything. Different conditions respond to different frequencies and intensities, much like how different radio stations broadcast on different channels. Athletes are using PEMF before workouts to boost performance and speed recovery—studies show it increases oxygen flow to muscles and helps them work more efficiently. In 2024, researchers began testing PEMF for PTSD, using artificial intelligence to customize treatments based on each person's unique electromagnetic signature. It's like the device learns to speak your body's electrical language. This personalized approach confirms what energy healers have long believed: true healing happens when we work with the body's own energy patterns rather than forcing change from the outside.

Red Light Therapy: Healing with Colored Light

Red light therapy uses specific colors of light—red and near-infrared—to trigger healing responses in your cells. Unlike UV light that can burn, these wavelengths are completely safe and don't produce heat. NASA originally studied this technology to help astronauts heal wounds in space, and now you can buy FDA-approved devices to use at home. The therapy works because our cells have tiny "receivers" (especially in the mitochondria, our cellular power plants) that absorb this light and use it to produce more energy. When cells have more energy, they can repair themselves faster and work better overall. The light can penetrate several inches into your body, which explains why the same basic technology can help with everything from reducing wrinkles to easing arthritis pain.

What makes red light therapy particularly interesting is that more isn't always better—there's a sweet spot where you get the best results. Too little won't do much, but too much can actually reduce the benefits. Studies show remarkable results for wound healing. In one trial, half of the pressure sores treated with red light healed completely, while none of the untreated sores improved at all. For skin health, people using red light masks for three months saw visible improvements in aging signs that lasted a full month after they stopped treatment. This tells us the therapy creates real, lasting changes in the skin rather than just temporary improvements. Some of the most exciting research involves shining red light into the brain through the skull or nose. Small studies have found improvements in memory and thinking in people with mild dementia after just eight weeks of treatment.

More and more clinics are now offering both PEMF and red light therapy, often using them together for better results. While they work through different mechanisms—PEMF uses magnetic fields while red light uses photons—they complement each other beautifully. PEMF can address whole-body electromagnetic imbalances while red light targets specific areas that need healing. Both therapies are remarkably safe, with virtually no side effects when used properly, making them excellent options for people who want to avoid medications or enhance their current treatments.

As we better understand how our bodies use electromagnetic energy, these tools are evolving from simple medical devices into sophisticated instruments that can actually communicate with our body's energy field, adjusting their output based on what we need moment to moment. We're moving toward a future where energy medicine works in perfect harmony with our body's natural electromagnetic blueprint, supporting healing in the gentlest and most effective way possible.

Beyond Treatment: Enhancement and Exploration

While most clinical applications of energy medicine focus on treating disorders, these technologies also hold potential for enhancing normal cognitive function and exploring altered states of consciousness. This raises both exciting possibilities and ethical concerns that merit careful consideration.

Research on cognitive enhancement through electromagnetic stimulation has shown potential benefits for various mental functions. Working memory—our ability to hold and manipulate information in mind—has been enhanced by tDCS applied to dorsolateral prefrontal cortex, a region crucial for working memory processes. These effects have been demonstrated in both healthy individuals and those with cognitive impairments, suggesting applications ranging from academic performance enhancement to rehabilitation after brain injury.

Attention represents another cognitive domain susceptible to electromagnetic enhancement. Various stimulation protocols have improved sustained attention, selective attention, and attention switching. For example, tDCS over right parietal cortex has enhanced vigilance performance in air traffic controllers during extended monitoring tasks. Such applications could be valuable in high-stakes environments where attentional lapses can have serious consequences, from surgery to aviation to military operations.

Learning and skill acquisition might also be accelerated through appropriate electromagnetic intervention. Stimulation during learning tasks has enhanced acquisition of various skills, from motor sequences to language vocabulary to mathematical procedures. The effects appear to work through multiple mechanisms, including enhanced neuroplasticity, improved attention to learning materials, and facilitation of the specific neural networks involved in the learned skill.

Perhaps most intriguing from a consciousness perspective, certain electromagnetic interventions appear capable of enhancing creativity—our ability to generate novel and useful ideas or combinations of ideas. For example, alpha-frequency tACS applied over frontal regions has increased performance on divergent thinking tasks, where participants must generate multiple creative uses for common objects. Other protocols have enhanced insight problem-solving—the "aha!" moment when a solution suddenly becomes clear after previous unsuccessful attempts.

The Future of Energy Medicine

The field of electromagnetic brain therapies is still relatively young, with many unanswered questions about optimal stimulation levels, individual differences in response, and long-term effects. But several trends suggest this field will continue to grow and evolve in exciting directions.

Increasing precision represents one clear trajectory. Advances in neuroimaging and computational modeling are enabling more precise targeting of stimulation to specific neural circuits, potentially improving efficacy and reducing side effects. Technologies like MRI-guided focused ultrasound can now create lesions in deep brain structures without invasive surgery, and similar approaches might eventually allow for targeted stimulation of deep structures as well. Optogenetics—using light to control genetically modified neurons—offers extraordinary precision in animal models, though translating this approach to humans faces significant ethical and technical challenges.

Personalization represents another important trend. There's growing recognition that optimal stimulation levels may vary between individuals and even within individuals over time. This has led to approaches that tailor treatment based on individual brain characteristics rather than applying one-size-fits-all protocols. Quantitative EEG analysis can identify specific oscillatory abnormalities in an individual, guiding personalized neurofeedback or stimulation protocols. Similarly, functional connectivity analyses using fMRI or MEG can identify network abnormalities unique to an individual, potentially allowing for more targeted intervention.

Perhaps most exciting is the development of closed-loop systems that combine stimulation with real-time monitoring of brain activity. These systems can modify stimulation based on the brain's moment-to-moment state, potentially enhancing effectiveness. For example, a closed-loop DBS system for epilepsy might monitor neural activity for patterns that predict an impending seizure and apply stimulation only when needed, rather than continuously. Similarly, a closed-loop transcranial stimulation system might adjust parameters based on real-time EEG feedback, ensuring that stimulation remains optimally aligned with the brain's natural rhythms.

The miniaturization and increased energy efficiency of electronic components is enabling the development of fully implantable systems for long-term use. Currently, most implanted neurostimulation devices (like those used for DBS) include external components for programming and power supply. Fully implantable systems would offer greater convenience, reduced infection risk, and potentially improve acceptance by patients.

Emerging technologies point toward even more sophisticated approaches. Virtual reality combined with biofeedback creates immersive environments that respond in real-time to a person's physiological state, providing powerful tools for resonance training. Imagine a virtual forest where the weather patterns directly reflect your brain and heart coherence, providing immediate feedback on your internal state while immersing you in a healing environment. Such technologies are already being developed for applications ranging from anxiety treatment to meditation training.

The integration of electromagnetic approaches with psychedelic-assisted therapy represents another frontier, using chemical interventions to facilitate shifts in resonance patterns while electromagnetic tools help stabilize beneficial changes. What's particularly interesting about psychedelics from a GRT perspective is their potential to facilitate lasting changes in resonance patterns. Studies of psilocybin-assisted therapy have shown enduring changes in brain network connectivity and oscillatory patterns months after a single session, particularly in individuals who report mystical-type experiences during their sessions.

Staying Grounded in Evidence

As with any emerging field, electromagnetic approaches to consciousness face both enthusiastic advocacy and skeptical critique. Navigating this landscape requires maintaining a balance between open-minded exploration and rigorous evidence evaluation. Some approaches have stronger empirical support than others, and even established techniques like TMS still have many unknowns regarding optimal parameters and mechanisms of action.

The field also attracts its share of exaggerated claims and pseudoscientific applications. Consumer devices claiming to enhance consciousness through electromagnetic stimulation often lack solid evidence, and some practitioners make claims far beyond what current science supports. Distinguishing promising approaches from premature hype requires both scientific literacy and a kind of disciplined open-mindedness—neither dismissing unconventional ideas out of hand nor accepting them without sufficient evidence. And when researching and buying specific products, definitely apply the time-tested principle of *caveat emptor* (let the buyer beware).

From a resonance perspective, this balance is particularly important. The recognition that consciousness is nothing but a nested hierarchy of resonant patterns opens exciting possibilities for understanding and modifying these patterns, but it doesn't in itself validate every claim about electromagnetic effects on consciousness. Maintaining scientific rigor while exploring these possibilities ensures that the field advances based on evidence rather than speculation.

Several principles can help maintain this balance. Mechanistic plausibility represents a crucial starting point—does the proposed intervention have a plausible mechanism for affecting neural oscillations based on known physics and neuroscience? Approaches that violate basic principles of physics or neurobiology should be viewed skeptically, while those with clear physical pathways deserve further investigation.

Dosimetry and targeting—determining whether an intervention delivers sufficient electromagnetic energy to the target neural tissue—represents another key consideration. Some consumer devices may not generate fields strong enough to significantly affect neural activity, especially when accounting for attenuation by the skull and surrounding tissues. Similarly, the spatial precision of targeting varies dramatically between approaches, from the highly focal effects of deep brain stimulation to the much more diffuse effects of consumer-grade tDCS devices.

Evidence quality and replication provide essential standards for evaluating claims. Single studies, especially those without appropriate control conditions, should be interpreted cautiously. Systematic reviews and meta-analyses that synthesize evidence across multiple studies offer more reliable guides to what interventions show consistent effects. And the gold standard remains independent replication by multiple research groups using rigorous methodologies.

They don't require dismissing unconventional ideas but rather evaluating them based on evidence rather than hype or tradition alone. This balanced approach allows the field to advance while avoiding both premature rejection of valuable innovations and uncritical acceptance of unfounded claims. It's worth noting that maintaining this balance doesn't mean demanding absolute certainty before pursuing research or even carefully implemented clinical applications. The history of medicine includes many effective treatments that were adopted before their mechanisms were fully understood. Aspirin was used for decades before its mechanism of action was discovered, and many psychiatric medications still have incompletely understood mechanisms despite clear evidence of efficacy. Similarly, electromagnetic approaches to consciousness might prove valuable even before we fully understand how they work, provided they show consistent evidence of benefit and acceptable safety profiles.

The key is to maintain both scientific rigor and some level of therapeutic optimism—using the best available evidence to guide practice while continuing to refine our understanding through ongoing research. This approach allows for responsible innovation while protecting against unfounded or exaggerated claims that might ultimately harm patients or discredit the field.

For individuals seeking to explore these approaches, whether for treatment of specific conditions or for consciousness exploration, critical thinking and informed consent are essential. Understanding what the evidence actually shows—rather than what marketing materials and anecdotes claim—allows for realistic expectations and appropriate risk-benefit assessment. Consulting with knowledgeable professionals, especially for more powerful or invasive interventions, provides an important safety check and personalization of approach. And staying informed about emerging research helps navigate a rapidly evolving field where today's cutting-edge approach may be superseded by tomorrow's innovation.

By maintaining this balanced approach—neither dismissing electromagnetic approaches out of hand nor accepting them uncritically—we can responsibly explore their potential for understanding and influencing consciousness while building a solid foundation of evidence to guide future development.

Beyond Either/Or

Another significant trend is the integration of electromagnetic approaches with other therapeutic modalities. Rather than viewing these approaches only as alternatives to pharmaceutical or psychological interventions, researchers increasingly recognize the potential for synergistic combinations. For example, transcranial stimulation combined with cognitive training appears more effective for cognitive enhancement than either approach alone. Similarly, neurofeedback combined with psychotherapy might accelerate emotional processing and behavioral change. And electromagnetic interventions might enhance the effects of psychedelic-assisted therapy by promoting the neural plasticity and network integration associated with therapeutic psychedelic experiences.

The shift from a chemical to an electromagnetic understanding doesn't invalidate the importance of neurotransmitters and other chemical factors. Rather, it suggests a more integrated view where chemical and electromagnetic aspects of brain function are seen as complementary rather than competing explanations. Neurotransmitters and neuromodulators influence the brain's oscillatory patterns, and oscillatory activity affects neurotransmitter release and receptor sensitivity. The two domains interact continuously in healthy brain function and can both go awry in pathological states.

What unites these diverse approaches—from ancient drumming circles to cutting-edge neurofeedback—is their direct targeting of the brain's electromagnetic patterns rather than its chemistry, which is a kind of intervention further down the chain of causality. Electromagnetic interventions work by modulating the oscillatory dynamics that, according to General Resonance Theory, comprise cognition and consciousness. This represents a paradigm shift in how we conceptualize and treat disorders of the brain and mind—moving from a primarily "prickles"-based understanding to one that incorporates the electromagnetic "goo" as central.

The impulse to explore, modify, and expand consciousness seems to be a fundamental human trait, expressed through whatever means are available in a given cultural and historical context. Our electromagnetic technologies represent the latest chapter in this ongoing exploration, offering new possibilities but connected to a very ancient human quest. These advances point toward a future where electromagnetic interventions play a central role in treating neurological and psychiatric conditions—probably not replacing pharmaceuticals or psychotherapy, but complementing them with approaches that more directly target brain and body oscillatory dynamics.

If consciousness involves specific patterns of resonant information integration, then directly modulating these patterns may offer more precise and effective approaches to disorders of consciousness than interventions that affect these patterns only indirectly. The growing use of electromagnetic approaches in clinical settings creates opportunities for more systematic research on consciousness. As large numbers of patients undergo these interventions for therapeutic purposes, researchers can collect standardized data on the effects of different stimulation parameters on different aspects of conscious experience. This creates a kind of systematic exploration of the parameter space of consciousness—mapping how specific changes in the brain's electromagnetic patterns correspond to specific changes in subjective experience.

Despite these exciting possibilities, electromagnetic medicine still faces significant challenges before it can fully realize its potential. Regulatory frameworks designed primarily for pharmaceutical interventions may not be optimally suited for evaluating electromagnetic therapies, potentially slowing their development and approval. The highly individualized nature of optimal stimulation parameters makes standard clinical trials more complicated than for one-size-fits-all drug interventions. And the complex, nonlinear dynamics of the brain's response to electromagnetic intervention require sophisticated modeling approaches still in development.

Additionally, as consumer devices for neurostimulation and neurofeedback become more widely available, there are concerns about their use without proper supervision or understanding. While democratizing access to these technologies could empower individuals to improve their own brain function, it also creates risks of ineffective or even harmful use without appropriate guidance.

Nevertheless, the trajectory is clear: electromagnetic approaches to brain health are moving from the margins to the mainstream, offering new options for treating conditions that have resisted conventional interventions. As our understanding of the brain's electromagnetic nature continues to deepen, these approaches are likely to become increasingly sophisticated, precise, and effective—potentially transforming our approach to mental health and our understanding of consciousness itself.

From Ancient Practices to Modern Technologies

While electromagnetic brain stimulation technologies represent cutting-edge approaches to manipulating consciousness, as mentioned previously they connect in interesting ways to ancient practices that have been used for consciousness alteration for thousands of years. This connection isn't coincidental but reflects the fundamental relationship between rhythmic stimulation and the oscillatory nature of consciousness—a relationship that our ancestors intuited long before we had the scientific framework to understand it.

Consider the practice of rhythmic drumming found in shamanic traditions worldwide. Anthropologists have documented remarkably similar practices across cultures separated by vast distances and time periods, from Siberian shamanism to Australian Aboriginal traditions to Amazonian healing practices. These traditions typically involve sustained drumming at specific frequencies, often around 4–7 beats per second—remarkably similar to the theta frequency range (4–8 Hz) associated with trance states in modern EEG research.

The neurological effects of such drumming have been documented in contemporary studies. Sustained exposure to rhythmic drumming can entrain neural oscillations to match the drum frequency, potentially inducing altered states of consciousness characterized by vivid imagery, decreased analytical thinking, and increased emotional or intuitive processing. From a resonance perspective, the drums create an external rhythm that entrains the brain's natural oscillations, shifting the overall pattern of neural synchrony and thereby altering conscious experience.

Chanting represents another ancient practice with similar effects. Vedic mantras in Hindu traditions, Buddhist chanting practices, Gregorian chant in Christian monasticism, and many other forms of sacred sound involve sustained rhythmic vocalization, often at frequencies that promote particular states of consciousness. Additionally, the specific acoustics of traditional chanting spaces—from Gothic cathedrals to Himalayan caves—often create resonance patterns that enhance particular frequencies, potentially amplifying their effects on consciousness.

Rhythmic movement and dance similarly appear across cultures as methods for altering consciousness. Sufi whirling, ecstatic dance practices in African traditions, and trance dances in many indigenous cultures all use repetitive movement to induce altered states. These practices synchronize sensory and motor systems, potentially creating coherent oscillatory patterns across neural networks. The addition of music and social synchronization (groups moving together) may further enhance these effects through mutual entrainment—multiple individuals synchronizing not just within their own nervous systems but with each other.

A comprehensive 2024 meta-analysis published in *Sports Medicine* found that structured dance of any genre was generally equal to and occasionally more effective than other types of physical activity interventions for improving psychological and cognitive outcomes including emotional well-being, depression, motivation, social cognition, and memory. Dance proved particularly effective for anxiety, depression, and health-related quality of life, especially in older adults, while in younger participants (under 16), dance was superior to other exercise in reducing somatization—the expression of psychological distress as physical symptoms.

From a resonance perspective, these findings make perfect sense: dance combines rhythmic movement that entrains neural oscillations, social synchronization that creates shared resonance between participants, musical entrainment that aligns brainwaves to therapeutic frequencies, and embodied expression that integrates cognitive, emotional, and motor systems. Unlike pharmaceutical interventions that target single neurotransmitter systems, dance engages multiple resonance mechanisms simultaneously—creating synchronized oscillations across motor, emotional, and social brain networks while fostering the kind of interpersonal coherence that may be fundamental to psychological well-being.

The contrast between dance interventions and antidepressant medications for treating depression is particularly striking. Meta-analyses consistently show that antidepressants produce an effect size below the threshold considered clinically meaningful. With this small effect size, the number needed to treat (NNT) is approximately 9, meaning that 8 out of 9 patients taking antidepressants receive no benefit beyond placebo. By contrast, dance interventions show an effect size twice as large as antidepressants and well above the clinical significance threshold. A recent BMJ network meta-analysis found that dancing produced the largest reductions in depression of all treatments studied, with SSRIs (a common type of antidepressant) by themselves being the least effective method examined.

This dramatic difference becomes even more significant when we consider that dance interventions come without the side effects, withdrawal syndromes, or unblinding problems that plague antidepressant trials. While pharmaceutical companies have spent billions promoting the now-debunked serotonin hypothesis of depression, the evidence suggests that getting people moving together in synchronized, rhythmic patterns—engaging their natural resonance mechanisms—may be far more effective than chemically manipulating neurotransmitter systems.

These traditional practices suggest an intuitive understanding of the relationship between rhythmic stimulation and consciousness—an understanding that modern electromagnetic technologies are now making more precise and controllable. The shamans, mystics, and contemplatives of the past may not have had EEG machines or transcranial stimulators, but they discovered through careful observation and experimentation that consciousness could be altered through rhythmic practices. Modern technologies can be seen as extensions of this ancient insight, offering more precise control but working through similar fundamental mechanisms.

But let's get back to the hippies, my people. When a hippie talks about "good vibrations" or getting "in tune" with someone, they are intuitively recognizing something that modern neuroscience is now confirming—that consciousness involves resonant patterns that can be influenced through rhythmic stimulation, and that these patterns can synchronize between individuals in powerful and healthy ways.

Electromagnetic approaches to consciousness, grounded in rigorous science but open to the wisdom of traditional practices, might represent the integration of these perspectives. Rather than dismissing traditional practices as superstition or modern technologies as soulless mechanism, we might recognize both as approaches to the same fundamental relationship between rhythm and consciousness—between cycles and conscious experience.

This integration is already happening in some contexts. Neurofeedback practitioners sometimes incorporate elements of meditation or breathwork into their protocols, recognizing that these traditional practices can enhance the effectiveness of technological intervention. Some researchers studying transcranial stimulation are investigating whether its effects might be enhanced by combining it with traditional rhythmic practices like music or movement. And contemplative traditions are increasingly open to scientific investigation of their practices, recognizing that modern understanding doesn't diminish their value but helps explain and potentially enhance their effects.

The impulse to explore, modify, and expand consciousness is very much a fundamental human trait, expressed through whatever means are available in a given cultural and historical context. Infants do it. Animals do it. We all do it. Our electromagnetic technologies represent just the latest chapter in this ongoing exploration, offering new possibilities but connected to a very ancient human quest.

Chapter Ten

Hippie Stew

The Dance of Consciousness

Over a decade ago, on the outskirts of a lush jungle overlooking the Pacific, I stood awkwardly at the edge of a wooden dance floor at the Kalani Retreat Center on the southern coast of the Big Island of Hawaii. My body felt stiff, my self-consciousness looming. The facilitator had explained that ecstatic dance was a free-form movement practice—no talking, no alcohol, no shoes, just moving however your body wanted to move. The concept terrified and intrigued me in equal measure. I had never danced as an adult without alcohol in my system, and usually quite a bit of it.

I watched as bodies of all shapes, sizes, and ages moved around the floor—some with graceful abandon, others with rhythmic precision, and many with what looked like chaotic spasms that somehow still maintained harmony with the evolving electronic beats. No one seemed to be watching anyone else; each dancer seemed immersed in their own experience while remaining mysteriously connected to the collective energy.

I closed my eyes, feeling the vibrations of the bass through the wooden slightly springy floor. My feet began to tap almost involuntarily, then my hips swayed slightly. Before I knew it, I was moving—tentatively at first, then with increasing freedom. Something was happening that transcended my intellectual understanding. The music wasn't just sound; it was becoming a physical force moving through my body. The dancers weren't just separate individuals; we were becoming a coherent field of motion, a synchronizing system. Stomping, twirling, jumping, and sometimes even hooting and hollering. Together.

I had found a living demonstration of resonance—of how disparate oscillating systems (in this case, human bodies) can synchronize through shared energy and rhythm. What I didn't know then was that this experience would provide not just personal liberation but a powerful metaphor for understanding consciousness and even the fundamental nature of reality itself. We weren't synchronized physically in other than a loose sense—ecstatic dance has no rules and no accepted "moves"—but we were synced up, we resonated together, in the moment, the enjoyment, the intentions that we shared. That's what mattered.

It was at Kalani Retreat Center, the birthplace of this "ecstatic dance" movement in the 2000s (Max Fathom, now living in Austin, Texas, was one of the key thought leaders and practitioners of this moving meditation dance practice), that has since spread globally, that I began to understand viscerally what my intellectual explorations had only hinted at: everything is vibration, everything is rhythm, everything is dance. Dance is the microcosm of the macrocosm of the microcosm…

Waves of Influence

In my early twenties (yes, again, going back to my Army years), Fritjof Capra's *The Tao of Physics* blew my mind wide open. Here was a physicist drawing explicit connections between modern quantum physics and Eastern mystical traditions. Both perspectives suggested that beneath the apparent solidity of matter lay vibrating patterns of energy—"it's all vibrations, man."

Capra wrote about how the quantum field theories developed in the early twentieth century had transformed our understanding of particles into vibrating patterns or rhythms of energy. This resonated deeply with ancient Eastern concepts like the Dance of Shiva in Hindu tradition—the cosmic dance that represents the rhythmic cycles of creation and destruction in the universe.

This convergence wasn't just poetic; it seemed to point toward something fundamental about reality. When I later encountered David Bohm's book *Wholeness and the Implicate Order*, which I've now read a few times and marked up profusely, these connections deepened further. Bohm proposed that the universe consists of a "holomovement"—a flowing, undivided wholeness where seemingly separate objects and entities are actually projections from deeper dimensions of reality, an "implicate order" that undergirds the explicate order that we live in.

What struck me most in Bohm's work was his insistence that consciousness isn't in any way separate from this wholeness. The observer and the observed are part of the same undivided reality. Yes, this bears many similarities to the "tropical panpsychism" ideas I developed in an earlier chapter. Here is that same truth, surfacing yet again.

By the time I encountered Lee Smolin's *The Trouble with Physics*, many years later as I continued my explorations into fundamental physics and cosmology, I was convinced that the traditional reductionist approach to physics had reached its limits. Smolin argued that string theory had led physics into a mathematical wilderness with little empirical foundation. What was needed was a more holistic approach that could bridge quantum mechanics and general relativity—perhaps one that could also accommodate consciousness.

What all these thinkers shared was a recognition that the problems of physics and the problems of consciousness might be intimately connected. You couldn't solve one without addressing the other. And this insight would eventually lead me to develop my own theories: first General Resonance Theory (GRT) as an approach to understanding consciousness, and later Quantum Ocean Wave theory (still very provisional at the time of this writing) as an extension into fundamental physics.

The Resonance Revolution

Standing on that dance floor in 2013, swaying to the music alongside dozens of other dancers, I experienced directly what I had been theorizing about: resonance as a fundamental organizing principle. Each dancer maintained their unique expression while simultaneously falling into sync with the collective rhythm—the many becoming one while remaining many.

This experience illuminated what would become the core principle of my General Resonance Theory: shared resonance in proximity is what allows micro-conscious entities to combine into macro-conscious entities. When neurons fire in synchrony—oscillating together at the same frequency—they create a larger, unified field of consciousness. Different levels of consciousness emerge through different patterns of resonance.

Just as dancers on that floor created a collective experience through shared rhythm, neurons and glia in the brain create conscious experiences through shared electrical oscillations that acquire their own even more complex structure above and beyond the firing patterns. And just as the dance floor contained many smaller clusters of synchronized movement within the larger whole, consciousness contains many nested levels of smaller conscious entities within the dominant conscious experience.

What I found remarkable was how the language of dance and rhythm provided such a natural vocabulary for describing these dynamics. We speak of being "in sync" or "finding our rhythm" or "resonating" with ideas or people. These aren't just metaphors; they point yet again to the fundamental vibratory nature of reality.

After that first experience with ecstatic dance, and as I attended more and more regularly, I began to see resonance everywhere—in human relationships, in ecological systems, in social movements, and most importantly, in the dynamics of consciousness itself. The hippies hadn't been wrong when they talked about "good vibrations" and "tuning in" with each other. They were intuiting something profound about the nature of reality. I am now firmly part of the "cult of ecstatic dance," as I affectionately call it, and I frequently attend this raucous party cum hippie church at Kalani near my house each Sunday. It is a wonder to behold. And even more wondrous to be part of that weekly hippie stew we co-create.

Schrödinger's Dream and the Ocean of Waves

This perspective led me eventually to revisit the work of Erwin Schrödinger, a Nobel Prize-winning physicist from Austria whose wave equation revolutionized quantum physics in the 1920s. I'd read about him going back as far as my teens, when I read John Gribbin's book, *In Search of Schrödinger's Cat*. While modern quantum mechanics retained Schrödinger's mathematics (his famous 1926 wave equation), it largely abandoned his philosophical vision. Toward the end of his life, in the early 1950s, Schrödinger insisted that "all is waves"—that reality consisted fundamentally of wave processes rather than particles.

With the centennial of Schrödinger's wave equation approaching in 2026, I've found myself drawn back to his original vision, now enhanced by a century of additional physics and a deeper understanding of consciousness. The result that I'm developing is called Quantum Ocean Wave theory—an extension of General Resonance Theory into fundamental physics. I write here tentatively, since this approach is still very much in progress and I want to be suitably humble about my efforts, as a non-physicist, to offer an extension of Schrödinger's system for the twenty-first century.

My approach reimagines the quantum vacuum as a dynamic, oscillatory medium—a vast ocean of vibration from which all physical phenomena emerge. Particles aren't separate "things" but are instead tight standing waves of energy in this infinite ocean. Forces aren't abstract fields but traveling waves of different frequencies of the same underlying energetic medium. The passage of time is simply the flex and flux of energy in this vast ocean.

What's remarkable about this framework is how naturally it unifies phenomena across vastly different scales. The same basic principles that govern quantum entanglement may also explain the shape and evolution of galactic structures. The mathematics reveals a smooth logarithmic relationship between frequency and velocity, with different forces emerging naturally at different frequency bands.

This approach resolves many long-standing puzzles in physics. Wave–particle duality isn't mysterious when everything is fundamentally waves—again: particles are just tightly confined standing waves of energy. Quantum entanglement makes sense when we understand it's maintained by what I call the "nu force"—waves in the quantum ocean that travel about 1,500 times faster than light (yes, that's very, very fast but still a finite speed). The mysterious boundary between quantum and classical behavior disappears when we see everything as wave patterns in the same medium, just at different scales and frequencies.

Most importantly, this framework provides a natural home for consciousness. If reality consists of waves in a quantum ocean, and the energy comprising these waves is inherently experiential, then consciousness can be understood as resonant patterns in this ocean—specifically, electromagnetic field patterns in the brain and body that integrate information through resonance, and of course in non-brain structures that also create nested electromagnetic field patterns, wherever those occur in nature.

Dancing the Cosmos into Being

The image of ecstatic dance provides a convenient metaphor for this worldview. Imagine the universe as a vast dance floor where patterns of energy move and groove across different scales. Elementary particles are the simplest dance moves—little vibratory patterns that repeat in consistent ways. When these particles "dance together" in synchronized patterns, they form atoms. Atoms dance together to form molecules, molecules form cells, cells form organisms, and so on up the scale.

Consciousness, in this view, is always present in each tiny form of matter and energy, but complex consciousness like that which we humans enjoy emerges when the dance becomes sufficiently complex and integrated through resonance across various scales. The brain is like a dance floor where billions of neurons perform synchronized routines, creating larger and larger coherent field patterns. The dominant consciousness we experience as humans is the highest level of this nested hierarchy of dances within dances within dances.

Yes, it's *dance* all the way down... man.

What's appealing about this metaphor for me is that it captures both the individual and collective aspects of reality. Just as each dancer on the floor maintains their unique expression while participating in the collective movement, each level of consciousness maintains its individuality while participating in higher levels of organization. This isn't just poetic metaphor (though it is that too); it reflects the mathematical reality described by Quantum Ocean Wave theory. The equations show how waves can combine to form stable patterns without losing their individual characteristics—"the many become one and are increased by one," as Whitehead put it succinctly (it's impossible to quote this passage too many times, since it's so foundational to Whitehead's system as well as GRT).

The hippies intuited this interconnected, vibratory nature of reality through their own direct experiences, often catalyzed by psychedelics, meditation, or practices like ecstatic dance. What they lacked was a rigorous theoretical framework to explain and extend these insights. That's what I hope to provide through the intellectual package of General Resonance Theory and Quantum Ocean Wave theory—a bridge between the intuitive wisdom of the counterculture, ancient wisdom traditions, and the mathematical precision of modern physics and cognitive science.

Today, when I return to the ecstatic dance floor at Kalani or elsewhere on the island (there are many venues for ecstatic dance now, including Sundari, Hawaiian Sanctuary, Garden Temple), I often move with a different understanding. Each gesture, each rhythm, each moment of synchrony with other dancers feels like a microcosm of the universe's fundamental processes. I'm not just moving my body; I'm participating in the same dynamic principles that govern quantum fields, galactic formations, and consciousness itself. My little ripples right here, right now, may in fact cause larger ripples *over there* and *over there*. Little ripples of love and joy floating and radiating outwards like energetic paisley patterns.

Nikola Tesla famously said: "If you want to find the secrets of the universe, think in terms of energy, frequency, and vibration." The hippies were right about the vibrations, just as Schrödinger was right about waves being fundamental. My hope is that bringing these insights together might help us develop a more complete understanding of consciousness and reality.

As I work on further articulating Quantum Ocean Wave theory (stay tuned for my next book and various peer-reviewed papers), extending the resonance principles I've developed for understanding consciousness into the domain of fundamental physics, I find myself returning again and again to the dance floor as both inspiration and testing ground. There's a direct line from that first shy shuffle at Kalani in 2013 to the equations describing waves in the quantum ocean.

The universe, as I've come to understand it, is indeed all waves, all the time, all the way down—a cosmic dance of vibrating patterns creating and dissolving across countless scales of space and time.

Chapter Eleven

Fields or Firings, Where the Heck is Consciousness?

Fair warning: This chapter gets a bit more technical than others in this book. Feel free to skim or skip it without losing the main thread of our story. But if you're curious about where exactly consciousness might live in the brain – and why that matters – stick with me.

Where exactly is consciousness in the brain? Or is it even "in" the brain at all, in the way we normally think about things being located somewhere?

Think of it this way: most scientists believe consciousness emerges from neurons firing and communicating with each other – billions of tiny electrical pulses jumping across synapses like messages passed between people in a massive game of telephone. This is the conventional view, what neuroscientists call the "spike code" approach. In this view, consciousness is basically the pattern of all those neural firings, like music emerging from an orchestra where each neuron is a musician playing their part.

But what if we've been looking in the wrong place? What if consciousness isn't in the neurons themselves, but in the invisible electromagnetic fields that surround and permeate the brain – fields created by neural activity but existing as something more, something that might be consciousness itself?

Here's something that might surprise you: your brain is essentially made of electromagnetic fields. I'm not being metaphorical here. Apart from the tiny nuclei at the center of atoms, literally everything in your brain – every thought, every memory, every sensation – is governed by electromagnetic forces. When neurons fire, when chemicals cross synapses, when you remember your first kiss or solve a math problem, it's all electromagnetic activity.

So the question isn't whether EM fields are involved in consciousness—they absolutely are. The question is whether the ambient fields generated by all this activity, the ones we can measure with EEG machines and other tools, are just noise—like the whistle on a steam locomotive that makes sound but doesn't help move the train—or whether they're actually doing something important for consciousness.

The Train Whistle Problem

Christof Koch, one of the world's leading consciousness researchers (and someone I've had the pleasure of interviewing a number of times), once compared brain oscillations to the sounds a beating heart makes. You can hear them with a stethoscope, they can tell you if something's wrong, but the body doesn't actually use those sounds for anything. They're just a byproduct.

For years, many neuroscientists have thought about the brain's electromagnetic fields the same way. Sure, we can measure them with EEG machines. Yes, they create pretty patterns that correlate with different mental states. But are they actually *doing* anything? Or are they just the brain's equivalent of a train whistle—making noise but not contributing to the actual work?

Koch was initially skeptical that these fields played any real role in consciousness. As he told me in a 2020 interview about his book, *The Feeling of Life Itself,* while it would be "foolish to categorically rule out any physical process," he wasn't convinced that specific frequencies of electromagnetic activity had specific functions in creating consciousness. But here's where things get interesting. My colleague Mostyn Jones and I (and many other scholars too) have been building a case that these electromagnetic fields aren't just along for the ride—they might actually be driving the train. Let me explain why.

First, think about what happens when doctors use techniques like transcranial magnetic stimulation (TMS) to treat depression or other conditions. They're literally using external magnetic fields to change how your brain works. If the brain's own electromagnetic fields were just noise—just that train whistle—then adding more electromagnetic energy from outside shouldn't change consciousness at all. It would be like trying to make a train go faster by making its whistle louder.

But these treatments work! They can lift depression, stop seizures, and create all sorts of changes in consciousness. As discussed in the last chapter, the science is very much still developing in this area, but early data are quite supportive of the conclusion that these therapeutics achieve real improvements in many psychiatric and psychological ailments. This strongly suggests that the brain's own electromagnetic fields must be doing something important, not just making noise.

Second, there's a basic principle in physics: everything in nature seems to work both ways. Push on something, and it pushes back. The idea that neural activity creates electromagnetic fields that then do absolutely nothing in return would be like having a one-way street in physics—it just doesn't happen in nature.

How Do You Have One Mind Instead of Billions?

Another puzzle that electromagnetic fields might solve. You have billions of neurons in your brain, each doing its own thing. Different areas process color, shape, motion, memory, emotion—all separately. Yet somehow, you experience one unified consciousness, not a chaotic collection of separate experiences.

Imagine you're seeing an old friend. One part of your brain recognizes the face, another processes the color of their shirt, another recalls memories, another generates emotional responses. But you don't experience these as separate events—you just see your friend, whole and complete, in one unified experience.

How does this happen? The conventional neural firing explanation struggles here. There's no single place in the brain where everything comes together, no "CEO neuron" that combines all the separate processing into one experience. It's like having an orchestra with no conductor, yet somehow producing a symphony instead of chaos. But electromagnetic fields might provide the answer. Unlike neural connections, which are like roads that only connect specific places, electromagnetic fields are continuous, like a fog that fills every space. They can reach across the entire brain instantly, potentially unifying all those separate processes into one conscious experience.

Here's where things get surprising. When neurons communicate through chemical signals across synapses, those signals travel at about 0.1 meters per second on average. Yes, you read that right. While some myelinated axons can conduct signals faster (up to 100 meters per second), the actual synaptic transmission—the jumping of signals from one neuron to another—is pretty slow when we consider the full span of the brain.

But electromagnetic field effects? According to groundbreaking calculations by Giulio Ruffini and colleagues, in an important 2020 paper, ephaptic field effects propagate at 47–57 kilometers per second at alpha frequencies (8–12 Hz) and an astounding 141–228 kilometers per second at gamma frequencies (30–100 Hz). Let's do the math: taking synaptic transmission at 0.1 m/s and ephaptic effects at, say, 50 km/s (50,000 m/s), we get a speed differential of about 500,000 times. Even using the more conservative comparison with the fastest axonal conduction (100 m/s), ephaptic effects are still about 500 to 2,000 times faster. But for actual synaptic transmission—the real bottleneck in neural communication—the difference is approximately 5,000 times faster, just as Ruffini's team calculated.

This isn't just a minor speed advantage—it's like comparing a text message sent by carrier pigeon to one sent by fiber optic cable. The implications are staggering. How could evolution possibly ignore such a massive efficiency gain?

The Science Revolution: From Train Whistles to Neural Orchestras

Over the past few years, the scientific understanding of ephaptic effects has undergone nothing short of a revolution. New experiments are revealing that these electromagnetic field effects aren't just real—they're fundamental to how our brains work.

The computational work by Ruffini's team has become so influential that leading neuroscientists are now incorporating these findings into their work. At the 2024 Tucson "Science of Consciousness" conference, I raised these speed calculations with MIT neuroscientist Earl Miller following his presentation. I saw him speak again at the 2025 Science of Consciousness conference in Barcelona and he has begun citing Ruffini's figures in his work and presentations. Miller, a titan in the field of working memory and cognitive control, recognizing these speed advantages represents another major shift in how mainstream neuroscience views electromagnetic field effects.

This is how scientific revolutions happen—not all at once, but through a cascade of researchers recognizing that the data can no longer be ignored. When someone of Miller's stature starts incorporating ephaptic speed advantages into their theoretical framework, it signals that the field is approaching a tipping point.

In 2024, a landmark paper in *Neuron* by Lee, Anastassiou, Koch, and colleagues further transformed how we think about ephaptic coupling. This wasn't just another incremental finding—it was a paradigm shift. The team demonstrated that different types of brain cells show distinct patterns of entrainment to electric fields, and this happens across multiple brain regions and species, including human neurons.

What makes this so significant? The paper positions ephaptic effects not as minor perturbations but as fundamental mechanisms providing an alternative, energy-efficient, non-synaptic paradigm for circuit coordination. In other words, electromagnetic fields aren't just influencing neural activity—they're a primary way brain circuits coordinate their activity.

Think of it this way: we used to believe neurons were like people passing messages by hand (synaptic transmission at roughly 0.1 m/s), and any electromagnetic effects were just background noise. Now we're discovering the brain also uses something like a sophisticated wireless network operating at speeds 5,000 times faster, with different types of neurons tuned to different electromagnetic "frequencies" for specialized functions.

Ruffini's team discovered something else fascinating: assemblies of neurons, when properly oriented, can function as "antennae" for ephaptic coupling. The brain isn't just generating electromagnetic fields as waste—it's structured to use them for communication. As they note in their paper, the brain appears to have evolved to take advantage of these field effects, with cortical columns and other structures such as the brain's ridges and grooves oriented to maximize ephaptic coupling.

The 5,000-fold speed advantage of ephaptic over synaptic transmission has profound implications:

1. Binding Problem Solution: The speed allows distant brain regions to synchronize within the narrow time windows required for conscious binding (typically 20–40 milliseconds for gamma-band synchrony).

2. Real-time Integration: Ephaptic effects can integrate information across the entire brain in under a millisecond, while synaptic transmission would take seconds.
3. Energy Efficiency: Ephaptic coupling requires no metabolic energy for transmission once the fields are generated, unlike the energy-intensive process of synaptic transmission.
4. Evolutionary Advantage: Any organism that could harness even a fraction of this speed advantage would have had enormous survival benefits.

More support comes from recent work by Dimitris Pinotsis at MIT, who works with Miller and has shown that ephaptic coupling may play a crucial role in memory formation. His 2023 research demonstrated that these field effects could allow memory networks to form in living brains—not just in laboratory slices but in actual, functioning neural tissue.

This suggests that memories might not be stored just in synaptic connections (like data on a hard drive) but also in patterns of electromagnetic fields (like holograms in space). With ephaptic effects operating 5,000 times faster than synaptic transmission, this could explain how memories can be retrieved almost instantaneously and why they seem to be distributed across the brain rather than stored in single locations.

The Clinical Revolution: From Theory to Treatment

Perhaps most excitingly, this new understanding is already influencing medical treatments. The recognition that electromagnetic fields are causally important—and operate at speeds thousands of times faster than synaptic transmission—explains why techniques like transcranial magnetic stimulation (TMS) and transcranial direct current stimulation (tDCS) can be so effective for conditions like depression, chronic pain, and epilepsy.

As Ruffini's team notes, if the brain naturally uses ephaptic mechanisms for communication at these remarkable speeds, then understanding these mechanisms should help us design better treatments. For instance, in conditions where ephaptic coupling is disrupted (potentially in disorders like schizophrenia or autism), therapies that restore normal field patterns might be more effective than drugs targeting specific neurotransmitters that operate at the much slower synaptic level.

Some of the most compelling evidence comes from Dominique Durand's lab at Case Western Reserve University. In 2019, they did something remarkable: they took slices of mouse hippocampus (the brain's memory center) and literally cut them in half, completely severing all neural connections. You'd think that would stop any communication between the two halves, right? Wrong. Slow waves of activity continued to propagate from one half to the other, jumping across the gap like electricity arcing through air. The only way this could happen is through electromagnetic field effects.

Even more amazingly, when they blocked these electromagnetic fields, the communication stopped. This proved that the fields weren't just noise—they were actually carrying information across the gap where no neural connections existed. And while these particular waves were slow (about 0.1 m/s), they demonstrated the principle that ephaptic effects can transmit information without any physical connections—a principle that operates at much higher speeds for the faster brain rhythms.

This has now been shown not just in brain slices but in living, anesthetized rats, and not just in the hippocampus but in the cortex too. The electromagnetic fields of the brain can make neurons fire without any direct connections between them, like one tuning fork causing another to vibrate through the air.

What this Means for Consciousness

If consciousness involves synchronized oscillations in neural populations—billions of neurons firing in coordinated rhythms—and if these synchronizations are mediated through electromagnetic fields operating 5,000 times faster than synaptic transmission, then these fields might literally be the substrate of consciousness, not just its byproduct.

Think of it this way: when large groups of neurons oscillate together at specific frequencies (like gamma waves at 30–100 Hz), they create coherent electromagnetic fields. According to Ruffini's calculations, these gamma-frequency fields propagate at over 140 kilometers per second—fast enough to coordinate activity across the entire brain in less than a millisecond. This could be the physical basis for the unity of consciousness.

I like to think of the relationship between the brain's anatomy and its electromagnetic fields like a tree. The brain's physical structure—the neurons, axons, and synapses—is like the trunk and branches. Solid, visible, structural. But the electromagnetic fields are like the leaves—more delicate, more dynamic, constantly rustling and changing with every breeze. Just as leaves grow from branches but aren't the same as branches, electromagnetic fields are produced by neural activity but aren't identical to it. They're more fine-grained, more sensitive, and—as we now know—capable of carrying information 5,000 times faster. These fleeting but incredibly swift fields might explain how our kaleidoscopic, ever-changing experience of consciousness emerges from the relatively fixed structure of the brain.

This debate has huge implications for artificial intelligence, the subject of later chapters in this book. Current AI systems, including large language models like ChatGPT or Claude, are based on the spike-code view—they simulate neural firing patterns digitally. But if consciousness actually requires specific electromagnetic field dynamics operating at speeds 5,000 times faster than synaptic transmission, then digital computers as we know them might be fundamentally incapable of consciousness, no matter how complex they become.

This doesn't mean machine consciousness is impossible, but it suggests we might need entirely different approaches—perhaps quantum computers or neuromorphic systems specifically designed to generate and sustain electromagnetic field patterns that can achieve the remarkable speeds we see in biological brains.

Christof Koch's Remarkable Change of Mind

Remember Christof Koch, the skeptic I mentioned earlier? In a remarkable intellectual journey that few scientists of his stature would make, he's gradually come to recognize that electromagnetic fields might indeed be causally important for consciousness.

This shift didn't happen overnight but emerged through careful consideration of mounting evidence. Work by researchers like Buzsáki, Freeman, and more recently Pinotsis, Miller, and Ruffini, as well as his own work discussed above, with Lee and Anastassiou, has solidified the case that EM fields are not mere epiphenomena but are in fact causally potent aspects of brain function operating at speeds that dwarf traditional neural communication.

Koch's intellectual courage in following the evidence led to our recent collaboration as guest co-editors on a major new "research topic" for *Frontiers in Human Neuroscience*: "Mapping the Resonome: Electromagnetic Field Patterns in Relation to Cognition and Consciousness." The "resonome"—the pattern of electromagnetic resonances in the brain—might be like a fingerprint of consciousness, containing far more information about mental states than neural anatomy alone, precisely because these fields can integrate information 5,000 times faster than synaptic mechanisms.

When one of the world's foremost consciousness researchers changes his mind based on evidence, and when computational neuroscientists like Earl Miller start incorporating these speed calculations into their theories, it suggests we're witnessing a genuine scientific revolution.

So where is consciousness? The evidence increasingly suggests it might not be in the neurons themselves but in the electromagnetic fields they generate—fields that resonate, interact, and create the unified experience we call consciousness at speeds 5,000 times faster than traditional neural communication.

We're not just biological computers processing information through slow synaptic transmission. We might be something more like electromagnetic symphonies, with consciousness emerging from the resonant harmony of billions of oscillating fields operating at near light-speed.

The science is beginning to catch up to the hippies' intuition. And that train whistle? Given that it operates 5,000 times faster than the train itself, it might actually be driving the whole show.

Chapter Twelve

Comparing Consciousness Theories

GRT, IIT, & Global Workspace Theory

In April 2024, I found myself back in Tucson, Arizona, for the 30th anniversary of The Science of Consciousness conference—a gathering that had shaped my thinking about consciousness very significantly over the years. The location, a resort in the foothills north of Tucson, is natural and remote enough that sometimes road runners and lizards can be seen passing by, checking out the visitors as we checked them out. Massive saguaro cactus grow majestically everywhere. It's a dangerous and exceedingly dry sort of beauty that is pretty much the opposite of Hawaii's lush and green beauty—nevertheless, I love both kinds of beauty.

I had regularly attended this conference starting about 2010 and have been to each conference since then. I've also been to the odd-year conferences, when it's held in an international location, a few times, including in Finland, Sicily (where I gave my first plenary talk), and in Barcelona in 2025. The 2024 conference was special—it was the 30th anniversary of the conference itself. That 30 years had seen massive changes in not only the science of consciousness, but also in the public acceptance of that science and the many, sometimes competing, theories that comprise that science. This thirty-year period also roughly marked the time that I had myself been seriously thinking about the nature of consciousness.

This year featured a panel that looked back on this 30-year renaissance and tried to coalesce some wisdom from what had transpired. My "academic daddy" Jonathan Schooler moderated a discussion at the end of the conference with the luminaries who had defined the field.

As I took my seat in the auditorium, I reflected on how this conference series had in some ways paralleled my own exploration of consciousness. At age 18, I'd devoured Douglas Hofstadter and Daniel Dennett's anthology *The Mind's I*, trying to wrap my head around the strange loops of self-reference, mathematics, and consciousness. At 19, Hofstadter's seminal *Gödel, Escher, Bach* had tickled my mind with its intricate fugues on minds and machines. And at age 23, I had been consumed happily in philosophical musings after reading Chalmers' seminal book, *The Conscious Mind*. Now, three decades later, I was watching many of these perennial themes play out on stage, between some of the key players in the field.

The panel assembled before us was remarkable. Stuart Hameroff, the conference's longtime organizer and anesthesiologist-turned-consciousness-theorist, sat next to David Chalmers, the philosopher who had coined the "hard problem" of consciousness. Christof Koch, the neuroscientist who has helped in many ways to develop Integrated Information Theory, was there alongside Susan Blackmore with her signature rainbow hair—a British philosopher known for her skeptical takes on consciousness and psi phenomena. Paavo Pylkkänen, the tall Finnish philosopher partial to Bohmian quantum mechanics, rounded out the group.

"The Science of Consciousness—30 Years On" read the title projected behind them. As I watched the luminaries discuss their perspectives, the conversation touched on all the major theories that had shaped the field. Chalmers spoke about how the hard problem of consciousness had evolved since he first introduced it at the inaugural TSC conference in 1994. Koch presented the case for Integrated Information Theory and its mathematical framework for measuring consciousness. Hameroff defended his Orchestrated Objective Reduction theory, noting recent discoveries in quantum biology that had vindicated some of his ideas. Blackmore, with her characteristic skepticism, questioned whether any of these theories truly bridged the explanatory gap between neural activity and subjective experience. Pylkkänen discussed how Bohmian mechanics and quantum field approaches might still offer insights into consciousness.

The panel discussion crystallized for me how each theory captured different aspects of consciousness while leaving others unexplained. Their exchange highlighted some of the central tensions in consciousness studies—between mathematical formalization and subjective experience, between quantum and classical approaches, between reductionist and holistic perspectives.

This diversity of views reinforced my sense that General Resonance Theory, a theory I had slowly crafted, with Schooler, after attending many previous TSC conferences, offered a framework that could integrate insights from many of these approaches.

The session concluded with reflections on where the field might be heading in the next thirty years. Despite their theoretical differences, all the panelists agreed that consciousness research had matured significantly since that first conference in 1994, evolving from a fringe topic to a legitimate scientific discipline. The conference's name highlighted this shift, changing from "Toward a Science of Consciousness" to "The Science of Consciousness." Had the field grown up and become an adult in this thirty-year span?

This maturity echoed my own journey. I'd come to see consciousness not as emerging from any single mechanism but as a manifestation of resonance principles operating across multiple scales.

The field had gone from a handful of researchers willing to tackle the study of consciousness scientifically to a thriving interdisciplinary field with multiple competing theories, and many conferences held around the world each year. The hard problem remained hard, but our understanding had deepened immeasurably. And, hey, we'd "solved" the "easy part" of the hard problem in our now well-known 2018 paper, "The Easy Part of the Hard Problem," which first introduced our GRT approach. (Yes, tongue firmly in cheek.)

Most importantly, the conference reinforced my conviction that advancing our understanding of consciousness requires not just rigorous theory and careful experimentation but also a certain humility about the limitations of any single perspective. The future likely lies in integration rather than conquest, synthesis rather than supremacy. Thirty years into this journey, many questions remain but our maps have become far more detailed, our questions more precise, and our appreciation of the depth of the problem more profound.

In the following pages I'm going to tease out some of the key differences between GRT and three of the more prominent theories of consciousness: Integrated Information Theory (IIT), developed primarily by psychiatrist Giulio Tononi (and his team at the University of Wisconsin) and neurophysiologist Christof Koch; the Global Neuronal Workspace theory developed by Stanislas Dehaene and his collaborators; and "predictive coding" as developed by Friston, Clark, Seth, and many others. Obviously, I'm partial to our own theory (GRT), and we crafted it specifically because it seemed that each of the prominent theories were missing key aspects. While I remain partial, I attempt to remain objective in the following discussion.

Integrated Information Theory: Consciousness as Intrinsic Integration

Integrated Information Theory represents one of the most mathematically rigorous and philosophically ambitious approaches to consciousness in contemporary science. At its core, IIT proposes that consciousness is identical with information integration in physical systems, which Tononi calls "integrated information" or "phi" (Φ). In later work, Tononi and Koch have clarified that what is being quantified and what is in fact consciousness is just causation. Information is a way of quantifying causal influence and thus consciousness in any system.

I've met Giulio a few times, and Christof many more times. I've interviewed them both, Christof a number of times now. They're both extremely impressive, intelligent, and driven. I was even inspired to call Tononi a "great man" after I interviewed him in San Diego in 2017, and shared oyster shooters with him and a small group at a restaurant that evening. He has a quiet intensity that is hard to resist being drawn in.

Integrated Information Theory begins with five axioms derived from phenomenology—intrinsic existence, composition, information, integration, and exclusion—and develops these into postulates about the physical systems that can support consciousness. In IIT, consciousness emerges from causal structures with specific topological properties; namely, those that integrate information in a way that creates a whole that is more than the sum of its parts.

There are many similarities between IIT and GRT. Both theories adopt a panpsychist orientation, suggesting that consciousness isn't confined to brains but exists across a spectrum of physical systems capable of integrating information in specific ways. IIT is more shy about its panpsychist implications, whereas GRT fully embraces them, for all of the reasons discussed in this book. Both approaches emphasize integration as the key to consciousness, recognizing that conscious experience is unified despite arising from diverse neural processes. And both seek to quantify consciousness, providing mathematical frameworks for measuring its presence and complexity.

However, there are also significant differences. IIT focuses on the causal architecture of systems—the way information flows and integrates within a network of elements, independent of the physical substrate. It's primarily a synchronic theory, examining consciousness at a particular moment rather than focusing on temporal dynamics or the particular substrate. GRT, in contrast, emphasizes the specific mechanisms of resonance for information integration and places greater emphasis on the oscillatory, dynamic nature of consciousness across time. GRT also suggests, as we've explored in detail in this book, that EM fields are the physical basis for consciousness. In other words, consciousness in GRT is very substrate-dependent—it's in the field, and in EM fields specifically.

This difference becomes particularly apparent in how the two theories approach the combination problem of consciousness. IIT addresses this problem through its "exclusion principle," which states that only one conscious entity can exist for any set of elements—the one with the maximum value of integrated information (Φ_{max}). This means that, in a human brain, if the entire cerebral cortex has the highest Φ, then it alone is conscious, and its constituent parts like individual cortical columns or neurons are not in any way conscious.

GRT takes a different approach, suggesting that consciousness exists at multiple scales simultaneously, with macro-conscious entities supervening on micro-conscious entities without extinguishing them. As Whitehead put it, "the many become one and are increased by one." This allows GRT to acknowledge nested hierarchies of consciousness—something IIT explicitly rejects through its exclusion principle. In IIT there can only be one conscious entity in any particular hierarchy, whereas GRT suggests a welter of nested conscious entities in any particular hierarchy. I contain multitudes. And so do you.

Another key difference concerns the role of time. IIT focuses on information integration within a system's causal architecture, largely independent of temporal dynamics. GRT, however, emphasizes the role of oscillatory processes unfolding over time—particularly the way neural systems achieve shared resonance through synchronized oscillations. This temporal dimension is essential to GRT's account of how micro-conscious entities combine into macro-conscious wholes.

Despite these differences, IIT and GRT may be seen as largely compatible at a deeper level. The integration that IIT quantifies with Φ may be physically realized through the resonance mechanisms that GRT emphasizes. The causal architecture that IIT describes may be implemented through the specific resonant relationships that GRT identifies. Each theory may be describing different aspects of the same underlying reality—IIT focusing on the informational/causal structure of consciousness, while GRT suggests the physical implementation through resonance dynamics in EM fields produced by the brain and body.

This potential compatibility has practical implications. For instance, GRT might help address one of IIT's limitations—its computational intractability for complex systems like human brains—by identifying key resonance mechanisms (quantified using what are known as "synchrony indexes") that indicate high levels of information integration without requiring exhaustive calculations of Φ values. The IIT team has developed a "perturbational complexity index" (PCI) that attempts a similar simplification for complex biological systems, but to my knowledge has not so far developed any synchrony index tools.

Global Workspace Theory: Consciousness as Global Availability

If IIT represents a primarily "intrinsic" theory of consciousness (defining it in terms of a system's internal causal structure), Global Workspace Theory (GWT) takes a more "extrinsic" approach, defining consciousness in terms of a specific functional role within cognitive systems. Originally proposed by Bernard Baars and further developed by researchers like Stanislas Dehaene, GWT views human consciousness as emerging when information becomes globally available to multiple cognitive processes through a kind of mental workspace.

The core metaphor in GWT is a theater where specialized, largely non-conscious processors compete for access to a global workspace analogous to a theater stage. When information gains access to this workspace, it becomes "broadcast" throughout the brain, allowing widespread access by attention systems, working memory, language, and other cognitive processes. This global broadcasting creates the reportable awareness that we associate with consciousness.

From a neuroscientific perspective, GWT focuses on specific neural dynamics associated with consciousness, particularly large-scale patterns of synchronized activity that may implement the global workspace. Dehaene and others have identified what they call "signatures of consciousness"—neural markers like the P3b wave in event-related potentials, gamma-band synchronization, and long-distance feedback signals from prefrontal to sensory areas—that correlate with conscious versus non-conscious processing.

There are significant areas of overlap between GWT and GRT. Both emphasize the role of large-scale synchronization in consciousness, recognizing that conscious awareness correlates with broader, more integrated patterns of neural activity. Both acknowledge the temporal dynamics of consciousness, with GWT describing a sequence from non-conscious processing to the "ignition" of global workspace activity. And both can account for the limited capacity of consciousness—in GWT through competition for the workspace, in GRT through constraints on what can be integrated through resonance at any given moment.

However, there are also important differences. GWT is primarily a cognitive theory focused on the functional role of consciousness in information processing, while GRT is more concerned with the physical basis and workings of consciousness itself. GWT is generally committed to a materialist/emergence view where consciousness is a product of specific brain functions, while GRT adopts a panpsychist perspective where consciousness is fundamental rather than emergent and becomes more complex through evolutionary processes. GWT typically restricts consciousness to cognitive systems capable of implementing a global workspace, while GRT suggests consciousness exists across a continuum of physical systems capable of achieving resonant integration, extending even to very simple systems (albeit very simply in such systems).

One of the most significant differences concerns the boundary of consciousness. GWT essentially equates consciousness with reportability—with information that has gained access to the global workspace and can therefore influence verbal reports, intentional actions, and other cognitive processes. This focus on reportability has led critics to suggest that GWT might be describing the neural basis of report rather than of consciousness itself—a distinction emphasized by consciousness researcher Ned Block through his concepts of "access consciousness" and "phenomenal consciousness." GRT takes a broader view, suggesting that resonant integration can create conscious states even in systems incapable of the kind of report emphasized by GWT.

Despite these differences, GWT and GRT may be seen as in many ways complementary. GWT provides a compelling account of how information becomes available for verbal report, intentional control, and reasoning—aspects of human consciousness that are undeniably important. GRT offers a broader framework for understanding how consciousness arises across diverse physical systems, how it relates to matter, and with human reportable consciousness representing just one particularly complex manifestation.

In practical terms, GWT's focus on empirically measurable "signatures of consciousness" provides valuable tools for studying human consciousness that complement GRT's broader theoretical framework. The neural synchronization that GWT identifies as a key marker of conscious processing aligns well with the resonance mechanisms emphasized by GRT. One can see GWT as offering a detailed account of a particular type of consciousness—globally broadcast, reportable human consciousness—within the broader spectrum of conscious phenomena addressed by GRT.

Predictive Processing: Consciousness as a Kind of Controlled Hallucination

A third framework that has gained prominence in recent years is the predictive processing approach, associated with researchers like Andy Clark, Karl Friston, and Anil Seth. This perspective views the brain as fundamentally a kind of prediction machine that constantly generates models of the world and itself, using sensory input primarily to correct these models rather than to build them from scratch.

In this framework, consciousness is sometimes characterized as a "controlled hallucination"—a brain-generated simulation constrained by sensory input but largely shaped by prior expectations and predictive models. The content of consciousness isn't a direct representation of external reality but a continuously updated prediction of that reality, constructed to minimize prediction error. As Kant observed centuries ago: we never truly know "the thing in itself." It's all approximations.

The predictive processing framework has interesting resonances with GRT, particularly in its emphasis on top-down influences in perception and cognition. Both approaches recognize that consciousness isn't simply a bottom-up process driven by sensory input but involves complex interactions between different levels of processing. Both acknowledge the active, constructive nature of conscious experience rather than viewing it as passive reception of information.

Where the approaches differ is in their metaphysical foundations and their emphasis on mechanism. Predictive processing is typically agnostic about the hard problem of consciousness, focusing instead on the functional architecture of cognitive systems and mostly avoiding the deeper philosophical questions. Seth likes to discuss what he calls "the real problem" of consciousness rather than the "hard problem." It doesn't directly address the question of why prediction-generating systems should be conscious at all. GRT, with its panpsychist foundation, offers a more direct engagement with this metaphysical question by suggesting that consciousness is an intrinsic aspect of all physical systems capable of resonant integration.

In terms of mechanism, predictive processing emphasizes hierarchical probabilistic inference implemented through precision-weighted minimization of prediction errors. This is the "predictive processing" that gives the approach its name. GRT focuses more specifically on resonance as the physical mechanism underlying integration of information. These aren't necessarily incompatible—the predictive architectures described in predictive processing could be physically implemented through the resonance mechanisms emphasized by GRT.

One particularly interesting point of convergence concerns the role of oscillations in neural function. Recent work in predictive processing has increasingly emphasized the role of neural oscillations in implementing prediction and error signaling, with different frequency bands potentially carrying different types of predictive information.

A recent paper by Ruben Laukkonen, Karl Friston, and Shamil Chandaria focuses on "sprays" (relatively slow neural mechanisms) and "fields" (the EM fields I've focused on in this book) as key mechanisms for broadcasting the predictive processing that is at the heart of their "beautiful loop" theory of consciousness. This proposal aligns well with GRT's emphasis on EM field resonance and neural synchrony as both fundamental to consciousness, though my recent work emphasizes just how much faster EM fields are than neural firing, and thus do far more of "the work" involved in cognition and consciousness.

Toward an Integrated Theory of Consciousness

As I reflect on these different theoretical approaches, what strikes me is not so much their contradictions as their complementarities. Each theory illuminates particular aspects of consciousness while leaving others in shadow. IIT provides a rigorous formalization of information integration but struggles with computational tractability. GWT offers compelling accounts of reportable awareness but may not capture the full spectrum of conscious phenomena. Predictive processing provides powerful computational models but doesn't directly address the hard problem. The following table summarizes some key comparisons.

GRT does not claim to resolve every challenge in consciousness studies, but it does suggest a unifying principle—resonance—that helps explain why the insights of IIT, GWT, and predictive processing each capture important aspects of consciousness while still leaving gaps. By focusing on resonance in nested electromagnetic fields, GRT offers a framework that can incorporate these diverse perspectives into a coherent whole.

The integration that IIT quantifies, the global availability that GWT describes, the predictive models of predictive processing, and the field-like properties emphasized by electromagnetic field theories might all be realized through the resonant EM field dynamics that GRT identifies as the basis for all types of consciousness.

	GRT	IIT	GWT	PP
Core Mechanism	Resonance in EM fields	Information integration Φ	Global broadcasting to workspace	Prediction error minimization
Substrate	Substrate-dependent (EM fields specifically)	Substrate-independent (causal architecture)	Brain-based neural networks	Agnostic
Metaphysics	Explicitly panpsychist	Implicitly panpsychist	Materialist/ emergentist	Agnostic on hard problem
Temporal Focus	Emphasizes oscillatory dynamics over time	Synchronic (moment-based)	Sequential ignition process	Continuous updating
Nested Consciousness	Multiple conscious entities can coexist	Exclusion principle—only one entity (Φ_{max})	Single global workspace	Not addressed
Scope of Consciousness	All physical systems capable of resonance	Systems with integrated information	Cognitive systems with global workspace	Systems minimizing prediction error
Reportability	Not required for consciousness	Not emphasized	Equates with consciousness	Not central
Hard Problem Approach	Direct—consciousness is fundamental	Mathematical formalization	Functional explanation	Avoids/"real problem" instead

A more complete understanding of consciousness may require integration across these levels. GRT, with its focus on a fundamental physical mechanism that spans many scales, and with its emphasis on quantifying and characterizing different forms of resonance, offers one path toward such integration. By identifying resonance as a universal principle that operates across levels of organization, GRT provides a conceptual bridge between different theoretical approaches.

A Test Case: Interpreting Split-Brain Phenomena

To illustrate how different theories might offer complementary insights, let's consider how they interpret a classic phenomenon in consciousness research: the split-brain patient. When the corpus callosum (the main connection between the brain's hemispheres) is severed, patients exhibit remarkable phenomena that raise profound questions about the unity of consciousness.

In these patients, information presented to one hemisphere remains inaccessible to the other. If a word is flashed to the left visual field (processed by the right hemisphere), the patient cannot verbally report seeing it (since language is typically left-hemisphere dominant) but can identify it by selecting an object with their left hand (controlled by the right hemisphere). This suggests some form of consciousness in the right hemisphere despite inability to verbally report.

Different theories interpret this phenomenon differently:

- Global Workspace Theory might suggest that each hemisphere in split brain patients implements its own global workspace, with consciousness arising when information is broadcast within that hemisphere-specific workspace. The inability to verbally report right-hemisphere information reflects the left hemisphere's lack of access to the right hemisphere's workspace.
- Integrated Information Theory would calculate Φ for various subsystems to determine which have the highest integrated information, and if the global phi value in a normal patient exceeds that of each hemisphere, then the nested hemispherical consciousnesses are extinguished and only the highest phi consciousness remains. Ditto for each layer of nesting. IIT then suggests that in split brain patients, where information flows have been dramatically reduced, each hemisphere constitutes a separate conscious entity with its own integrated information, potentially explaining the apparent independence of awareness in each hemisphere.
- Predictive processing might focus on how each hemisphere generates its own predictive models based on the information available to it, with these models remaining distinct due to the severed corpus callosum. The divergence in behavior reflects these separate predictive models.
- General Resonance Theory, as an electromagnetic field theory of consciousness, suggests that severing the corpus callosum disrupts the major neural pathways between hemispheres while leaving the brain's electromagnetic field dynamics partially intact. Unlike purely neuroanatomical theories, GRT recognizes that electromagnetic fields can propagate to some degree across brain tissue regardless of anatomical

connectivity, potentially allowing some forms of inter-hemispheric resonance to persist despite callosotomy. This perspective aligns with emerging evidence that split-brain patients retain access to certain types of cross-hemispheric information, suggesting that consciousness operates at least in part through field-based mechanisms that transcend discrete neural connections. Callosotomy results in a modified but still partially integrated conscious experience where field-based resonance provides a substrate for an ongoing unified consciousness, of sorts, across both hemispheres.

Each of these interpretations captures something important about split-brain phenomena. GWT highlights the role of reportability, IIT emphasizes information integration, predictive processing focuses on model generation, and GRT suggests the physical mechanism of resonance may continue even after surgery.

The Future of Consciousness Theory

Conceptual integration will identify commonalities between theories that currently use different terminology and formalisms. For instance, the information integration emphasized by IIT, the global availability central to GWT, and the resonance highlighted by GRT might be recognized as different aspects of the same underlying process.

Methodological integration will combine diverse approaches to studying consciousness, from formal mathematical analysis to neuroimaging to neurophenomenology to comparative research across species. No single method captures all aspects of consciousness, but an integrated methodological toolkit might collectively address the full spectrum of conscious phenomena.

Disciplinary integration will bring together insights from neuroscience, psychology, philosophy, physics, biology, computer science, and other fields, recognizing that consciousness intersects with all these domains. The artificial boundaries between disciplines often obscure connections that become apparent when viewed from a broader perspective.

This integrative approach aligns with what philosopher Thomas Nagel has called "the view from nowhere"—the aspiration to integrate subjective and objective perspectives rather than privileging either. In the context of consciousness theory, this means honoring both the first-person, phenomenological dimensions of consciousness and the third-person, scientific approaches to studying it.

General Resonance Theory contributes to this integration by identifying a universal principle—resonance in EM field systems—that operates across levels of organization from quantum to cosmic. By focusing on this principle, GRT provides a potential bridge between subjective and objective approaches to consciousness, between different theoretical frameworks, and between scientific and contemplative understandings of mind.

Chapter Thirteen

What the Hippies Got Wrong

In the summer of 1983, the Love Family fell apart in a dramatic schism. I was twelve years old and had been living with the Love Family for less than a year. I watched as the adults I'd been taught to trust tore each other apart over money, power, sex, and conflicting visions of utopia. Love Israel, the charismatic leader who preached unity and divine consciousness, had been exposed for financial improprieties and sexual exploitation of his position. The same man who taught us that "we are all one" had created a rigidly hierarchical system where he lived in relative luxury, servants on hand at all times, and had sexual access to almost any woman—or girl—that he wanted, while many families struggled in poverty and extremely basic living conditions.

Decades later, reflecting on that rupture, it's not hard to see it as emblematic of a larger pattern—what I've come to call the "hippie paradox." The counterculture that gave us profound insights about consciousness, the nature of reality, and the interconnectedness of all things, also produced staggering examples of hypocrisy, irresponsibility, and spiritual bypassing. The very people who understood that "it's all about vibrations, man" often failed spectacularly at the basic vibrations of human decency and responsibility.

Love, Power, and Abuse: The Love Family

The Love Family was one of the largest and best-known "communes" to emerge from the counterculture in Washington state during the late 1960s and 1970s. At its maximum, it included 350–400 people spread over half a dozen communities. The two main communities were in Seattle's Queen Anne neighborhood and a ranch near Arlington, about an hour north of Seattle.

Founded by Paul Erdmann, a former television salesman who renamed himself Love Israel after his own psychedelic transformations, the community was built around the principle that "We are all one," "Love is eternal," and "God is Love," among other admonitions. Members adopted "Israel" as their surname and after fully committing to the Family took on "virtue names" like Harmony, Truth, Logic, Integrity, Compassion, and Serious. They lived communally, shared resources, and practiced a unique blend of Christianity, Judaism, mysticism, and psychedelic spirituality.

Understanding Ruth Israel, who lived in the Love Family for 26 years, described what drew people to the community: "The thing I noticed the most from reading those [stories of how people joined] was that people were young. They were looking for some way to work together and live together… there were a lot of young beautiful people and it was a really happening thing… it had some spiritual overtones, meditation." The community offered a compelling blend of spiritual practices: "What we did is every morning we meditated… we had a lot of Jews that joined the family too… we [celebrated] Passover, we're wearing robes, we're living the Old Testament… along with Eastern meditation, living, sitting in lotus position, chanting."

On the surface, it seemed idyllic—a utopian alternative to the materialistic and violent mainstream culture. Music, dance, self-reliance, gardening, spiritual practice, peace and love, and communal living were the beautiful aspects of this new/old experiment in living. Members believed they were creating a more spiritual, connected way of life. A different way of living and being in the world. A kind of heaven on Earth. Love, the leader, liked to talk about how the Family was the new incarnation of the twelve tribes of Israel.

I personally joined the commune after my brother joined. He and my mom and stepdad had met some members of the commune in Mexico, and later visited with the Love Family in Seattle, while roadtripping through North America. My brother decided he wanted more stability than he was getting with my mom or my dad and he asked if he could stay at the commune. My mother, like a good liberal hippie mom, said yes. I was staying with my dad in California and we heard about this and decided to hitchhike up (yes, the whole way) from Northern California to Seattle. Once I'd stayed a little while at the Front Door Inn, a free lodging house that the Love Family ran as a public service and an entry point for recruiting new members, I decided I really liked it too and I asked if I could also stay. My dad of course said yes (I know, this is rather strange by today's standards but I guess it wasn't that weird back then) and I was assigned to a household out at the family's Ranch near Arlington, about an hour north of Seattle. That household was run by Integrity and Confidence, a couple with a bunch of kids living in yurts on the shore of a beautiful lake with a dock and a canoe that I loved to take out. I remember that the lake had small leeches in it so you had to be careful to find them when you got out, or just keep moving and they wouldn't get on you. I actually loved my time at the Ranch and the Love Family more generally, I don't recall any issues that I had myself (but then again, I was a ten-year-old boy and faced different circumstances). I was sad when I was told by my temporary family that I would have to leave the Love Family because there had been a massive shakeup in the commune—the big 1983 schism.

The unvarnished reality for kids and adults alike who lived there more permanently, as Rachel Israel describes in her memoir *Counterculture Crossover: Growing Up in the Love Family*, was far more troubling. The insular, us-versus-them mentality developed quickly. As Understanding Ruth Israel observed: "We would call [outsiders] more worldly than us. And because I remember I was stirring granola with our cotton dresses and our hair tied back when the door opened. These beautiful women with low cut synthetic material dresses came in… and Jody said, 'We're going to watch our men, right? They're going to take our men.'"

The community operated with rigid male hierarchies disguised as spiritual designations. Understanding told me: "It was pretty clear [in terms of] male roles and female roles… Pretty much women were domestic and men were digging up the earth and chopping the wood." Love lived in comfortable houses (yes, plural) with his multiple wives, while ordinary families crowded into military surplus tents and flimsy yurts, sometimes with minimal protection from the harsh Pacific Northwest winters. Many struggled with inadequate food, clothing, and healthcare while Love and his inner circle grew fat.

The financial structure was particularly exploitative. Understanding explained the process: "After you got into the family after baptism… you had to sign a paper that everything you own belonged to the love family… you signed away everything… and it was called last will and testament you actually signed it." New members were required to donate all their possessions and money to the community—supposedly to transcend material attachment. But these resources were controlled entirely by Love and his inner circle—all men. Members worked in Family businesses or at outside jobs, but their earnings went straight to the communal coffers, which they had no say in managing. When questioned about finances, Love would dismiss concerns as evidence of insufficient spiritual development.

Even more disturbing was the sexual exploitation. Under the guise of spirituality, Love claimed sexual access to, eventually, all women in the community. Young women were often "guided" toward relationships with older men in the inner circle, including Love. The notion that "we are all one family" was used to blur boundaries, sometimes even with minors. The community's isolation and rejection of mainstream social norms created an environment where abuse could and did flourish unchecked.

The community's deliberate rejection of conventional structures extended to refusing to acknowledge peoples' age. Members were expected to say "we are all eternal" rather than give their actual ages, which created dangerous situations where adults couldn't properly protect children. Understanding witnessed this policy's foreseeable consequences firsthand:

"And so one of [the new kids] was sent to me and her name in the family was [blank]… and one night I found a man in her bed and I knew, I thought, maybe she's 15 or 16… so I mentioned it to Love [as a problem]… and then she came to me [later] and she said, 'Hey, my mom's coming for my birthday and I'm pregnant.' And so I knew. Her mom came and went and I had a friendship with her. So she said, 'Will you tell my mom I'm pregnant?' I said, 'your mom's going to be a grandmother. She's going to love you.' She said, 'No, she isn't. I'm going to be 13.' And yeah, I changed [my views] that second. we are on the wrong path. We need ages. This is abuse."

I know personally of at least one case where Love himself was the sexual abuser of an underage girl, again I will leave names out of this account.

The Love Family also used psychedelic drugs as tools of control and conformity. Understanding described how "We would have meetings where everybody took LSD and if you were with the kids like me, I didn't have to take LSD… My twin sister joined and she didn't want to do it and that caused a real problem. They saw you as a traitor." What had begun as consciousness expansion became coercive groupthink.

Children in the Love Family lived an especially strange existence. We were taught that our biological families were less important than our spiritual one. The discipline was strict and often arbitrary—we were told how to dress, how to speak, how to behave, all according to Love's ever-changing whims. Education was minimal and focused primarily on Love's teachings rather than practical knowledge that might help us function in the outside world.

For Understanding, who had sacrificed everything to join the commune—her marriage to Claudette Baker Rockefeller (a second cousin of Nelson Rockefeller), custody of her two daughters, and her mainstream life—the revelation about child abuse became a turning point: "So, that changed me because I went, 'This is the wrong path. I don't know what I'm going to do, but I decided I got to stand up for my values. I cannot just follow Love and these brotherhood. This is wrong.' so that was one case of abuse."

The Love Family represents one of the most tragic examples of how beautiful ideals about universal love and spiritual community could be corrupted by unchecked power, financial exploitation, and the dangerous rejection of basic protective structures that mainstream society had developed for good reasons. The very people who understood that "it's all about vibrations, man" often failed spectacularly at the basic vibrations of human decency and responsibility.

The Great Man's Shadow: Alan Watts

This pattern of deep spiritual insight coupled with personal contradiction was embodied in similar but less extreme ways in Alan Watts, the British philosopher who became the counterculture's most eloquent interpreter of Eastern wisdom to Western minds.

Born in 1915 in England, Watts showed an early interest in Eastern philosophy and Buddhism. By the time he was a teenager, he was already writing on spiritual subjects and had begun practicing meditation. After moving to America in the 1930s, he went on to become one of the most influential voices introducing Zen Buddhism, Taoism, and Hindu philosophy to Western audiences.

Watts possessed a remarkable gift for explaining complex Eastern concepts in accessible, engaging terms. His books, including *The Way of Zen, The Book: On the Taboo Against Knowing Who You Are,* and *Nature, Man and Woman,* became essential reading for the counterculture. His radio broadcasts and lectures drew thousands of listeners and attendees, and his ideas profoundly influenced musicians, artists, and thinkers throughout the 1960s and early 1970s.

What made Watts particularly appealing was his rejection of spiritual pretension. He didn't present himself as a guru or saint but as a "philosophical entertainer" who delighted in exploring the paradoxes of existence. He could speak for hours about the illusion of the separate self, the dance of consciousness, and the fundamental unity of existence. His books and lectures helped shape my parents' worldview and, through him, my own understanding of consciousness. His ideas and voice seem to have grown only more prominent after his death.

Yet behind this public persona lay a deeply contradictory private life. Watts married three times and fathered seven children, but by all accounts was often absent from their lives. His first marriage, to Eleanor Everett, produced two daughters but ended when Watts began an affair with Dorothy DeWitt, whom he later married. With Dorothy, he had five more children but was reportedly an inconsistent and distant father. His third marriage, to Mary Jane Yates King (known as "Jano"), came after meeting her during a lecture tour in New York.

Watts' drinking became increasingly problematic over his lifetime. According to biographers, by the end of his life, he was consuming a bottle of vodka daily. His heavy smoking, combined with alcohol abuse, damaged his health severely. He died in his sleep at age 58, his liver damaged by decades of heavy drinking, having spent his final years in an alcoholic haze that seemed to contradict much of what he taught about awareness, divinity, and clarity.

The contradiction is striking: here was a man who spoke eloquently about being present in the moment who was rarely present for his own family. He preached about the interconnectedness of all beings while maintaining careful distance from those closest to him. He advocated mindful awareness while increasingly numbing himself with alcohol.

According to a relatively unflattering biography (*Genuine Fake: A Biography of Alan Watts*) written by Monica Furlong, Watts' first wife Eleanor wrote an anguished letter to a bishop as their marriage was breaking up, describing Watts' interest in "painfully kinky sex." His second wife, Dorothy, reportedly wanted a more conventional life than Watts was willing to live. Though he maintained relationships with many women, including a long-term connection with Jean Burden (who inspired his book *Nature, Man and Woman*), these relationships were often characterized by distance and detachment rather than genuine intimacy. Furlong's title is intentionally ironic: Watts did indeed consider himself a kind of "genuine fake."

Watts was at least honest about his contradictions. In his autobiography, *In My Own Way* (the title itself a playful double entendre), he described himself as "a sedentary and contemplative character, an intellectual, a Brahmin, a mystic and also somewhat of a disreputable epicurean who has three wives, seven children and five grandchildren." He acknowledged what he liked to describe as his "irreducible rascality"—a phrase that served as both confession and excuse for his sometimes profound personal failings. We are all rascals at some deep level, he claimed. This framing was both honest but also an ever-present excuse for Watts acting in ways that others didn't always appreciate.

Critics suggest that Watts' lifestyle was driven partly by financial necessity—his three marriages resulted in alimony payments that required him to maintain a punishing lecture and writing schedule. But this explanation doesn't fully account for the contradiction between his insights and his choices. Watts understood intellectually that spiritual liberation wasn't about escaping responsibility but embracing it fully—yet he seemed unable or unwilling to apply this understanding to his own life.

His self-awareness was both his strength and his weakness—it allowed him to speak authentically about the human condition, but it also provided a convenient excuse for never seriously attempting to align his behavior with his teachings. Watts could discuss the illusion of ego while indulging his own ego's desires. He could explain the importance of compassion while failing to meet the needs of those who depended on him. He could describe the limitations of seeking fulfillment through pleasure while continually pursuing his own pleasures at others' expense.

This discrepancy between teaching and living creates a difficult question for those who value Watts' insights: can we separate the wisdom from the flawed messenger? Should we? The answers aren't simple, but they go to the heart of the hippie paradox—the gap between cosmic awareness and ethical responsibility that characterized so much of the counterculture.

Personally, I have no problem revering Watts as a great teacher, and even calling myself (playfully, of course) a *Wattsist*. He is my favorite philosopher and spiritual teacher! That said, I don't expect perfection from my spiritual leaders or teachers, and I, as we all do, also have some of the irreducible rascality that Watts liked to point out. And yet. And yet. Clearly, Watts could have been a much better father, spouse, and friend. Like most charismatic and highly intelligent people, he seemed to be more than a little corrupted by his own power, perhaps more than a little bit a narcissist. This, too, is part of the human condition. And one of Watts' primary admonitions was this: be a complete human being. He was definitely that.

My Father's Contradictions

This same paradox lived in my father, Christopher Hunt—or Caras E Esta, as he preferred to be called in the later years of his life. Born Christopher Robin Hadley in Devon, England, in 1945, my father was adopted as an infant by Ernest and Dorothy Hunt. Like Watts, he showed an early interest in spirituality and Eastern philosophy, though without Watts' formal education or discipline. Where Watts had studied comparative religion at seminary, my father was largely self-taught after high school, a kind of street philosopher (or *philosophe*) who absorbed ideas from books, countless conversations, TV, and his own psychedelic experiences and travels.

He was a man of enormous contradictions. In many ways, he embodied the hippie archetype—long-haired, free-spirited, disdainful of conventional society, passionate about alternative consciousness and spiritual exploration. He would speak for hours about cosmic unity, environmental stewardship, and the need for a more compassionate world. *The man could talk.* When he was awake he was talking, cigarette in his mouth, cup of coffee in his hand. Only when he slept would his mouth stop moving. He had genuine insights about consciousness and humanity's place in the cosmos that influenced my own philosophical development greatly over the years. He planted seeds in me that took many years to grow, but grow they did, as I recognize now from the vantage point of my own middle age.

Photo: My father and I in Pai, Thailand, in 2017.

But those insights existed alongside staggering blind spots and inconsistencies. He could speak passionately about personal empowerment and the rights of the downtrodden while simultaneously admiring dictators like Xi Jinping and Napoleon. This particular contradiction always baffled me—how could someone who claimed to value individual freedom and expression admire authoritarian leaders who suppressed those very qualities? When challenged on this point, he would become defensive and dismissive, unable to acknowledge the glaring inconsistency.

Similarly, he preached environmental consciousness and criticized modern society's destruction of the natural world, yet did little to reduce his own environmental impact. He flew frequently, maintained a significant personal carbon footprint, and when I once questioned him about this disconnect, he became angry rather than engaging with the critique. His environmental concern, it seemed, was more theoretical than practical—a position to be advocated but not necessarily lived.

Most significantly, he advocated for peace, love, and connection while abandoning his actual responsibilities to those closest to him. As a young husband and father in Cornwall, England, my father had briefly supported our family through his work as a potter. But that conventional path didn't satisfy him. When I was just a toddler and my brother a couple of years older (neither of my sisters were yet born), he convinced my mother to pull up stakes and took us all to India on a hippie spiritual quest.

That journey marked the beginning of his effective abandonment of our family. After returning from India, he increasingly withdrew from family responsibilities, eventually leaving my mother to raise three children (once my sister joined the world) on welfare in England while he traveled the world seeking entertainment and enlightenment. This wasn't a clean break or an honest decision to end the relationship—it was a gradual fading away, a shirking of responsibility disguised as necessity, to earn money, to achieve things in the world—anything, it seemed, to avoid staying in England with his young family.

And as I learned recently from my mother, this abandonment after the India trip was not the first time he had disappeared and left my mother in the lurch with a child in her belly, or on her hip, or both. It was kind of a pattern (it pains me to write this, but it's the truth).

The financial abandonment was almost complete. Child support was largely non-existent. My mother struggled to feed and clothe us on minimal government assistance while my father pursued his adventures, apparently unbothered by the material realities his choices created for his wife and children.

His drug use exemplified the hippie paradox perfectly. Cannabis and psychedelics had genuinely opened his mind to expanded states of consciousness, leading to insights about the nature of reality that I still find valuable. Most of my adult life that I knew him he wasn't much of a drug user other than the ubiquitous coffee and cigarettes—or the Thai cheroots (rolled leaf cigarettes) that he loved. By the end of his life, however, he was (I learned after his death, from the autopsy report) using cocaine and heroin, chasing a chemical solace in Thailand at age 73, and which drugs were very likely the proximate cause of his death.

I learned the news of his death just a few days after I had seen him in Georgetown, Malaysia, where I was traveling with my then partner, Olivia. He had been very difficult to be around on that trip, angry about everything. We had a number of blowups. It was an unpleasant visit. We had left Georgetown on bittersweet terms, but I got a great photo of him and a nice little movie of him on his scooter, with a smile.

Photo: My father in Georgetown, Malaysia, the last time I saw him, in 2018 a few days before his death in Thailand.

When I landed in Hanoi on my way back to home in Hawaii, I saw a number of messages on my phone and knew right away, without even reading them, that he had died. I changed my plans and bought a ticket back to Chiangmai, then took the van shuttle from Chiangmai to Pai, around 600 mountain bends, arranged a wake for him, and did my best to wind up his affairs in that small hippie town where he had been living for three years, mostly on money he had inherited from his father.

The pattern was consistent: genuine insight without integration, cosmic awareness without ethical responsibility. My father saw himself as a modern-day shaman and philosopher, planting seeds of a new consciousness in everyone he met. And perhaps he was, in his way. He certainly touched many lives and sparked genuine awakening in some, including me. But he was also a deadbeat dad who could talk for hours about the unity of all things while remaining profoundly disconnected from his own family's needs.

His sense of himself as a mystical figure extended to elaborate fantasies about his identity. In his later years, he claimed to be Caras E Esta, "keeper of the west wind," a title supposedly bestowed by an ancient pre-druidic order called the Kierna Kai. He claimed to speak their language fluently and to possess magical powers including weather control and divination, though he conveniently chose not to use his powers of immortality because it would mean watching loved ones die.

This story—and it was most likely just a story, though I remain at least a little open to it having some semblance of truth—seemed to be designed to elevate him above ordinary human concerns. Why worry about child support when you're a mystical keeper of ancient wisdom? Why bother with consistent employment when you're destined for greater spiritual work? Why honor commitments when you're operating on a higher plane of existence?

My father's story, like those of Love Israel and Alan Watts, illustrates how easily spiritual insight can become spiritual bypass—a way of avoiding rather than embracing the full spectrum of human experience and responsibility. The cosmic perspectives gained through meditation, psychedelics, or philosophical exploration can become escape hatches from the messy, difficult work of being fully human, and being fully in integrity with our own higher values.

The Hippie Paradox

This pattern—profound insight coupled with profound irresponsibility—appears throughout hippie culture. It's the ability to perceive deep truths about consciousness while failing spectacularly at basic human responsibilities. The very openness that allowed hippies to discover genuine insights about reality also made them vulnerable to almost every form of self-deception and excuse-making.

Consider the common hippie traits:

- Flakiness disguised as spontaneity: The emphasis on "going with the flow" often meant breaking commitments and abandoning responsibilities whenever something more interesting came along.
- Laziness rationalized as anti-materialism: The rejection of "straight" jobs and conventional success too often became an

excuse for refusing to contribute meaningfully to society or support one's dependants.

- Emotional and spiritual bypassing through flowery language: Real conflicts and hurts were dismissed with phrases like "it's all love" or "everything happens for a reason," preventing genuine accountability or healing.
- Anti-authoritarianism that enabled new tyrannies: The rejection of traditional hierarchies often created power vacuums filled by charismatic manipulators like Love Israel.

The hippies were right that conventional society was riddled with problems—materialism, militarism, conformity, disconnection from nature and spirit. But their alternatives often created new problems perhaps just as severe. They threw out the baby with the bathwater, rejecting not just societal constraints but also the wisdom that some structures and commitments serve essential human needs.

What makes this particularly frustrating is that many hippie leaders, like Watts, were fully aware of their contradictions—and their irreducible rascality. This self-awareness without self-improvement became another kind of spiritual bypass. By acknowledging their flaws with a wink and a mystical explanation, they avoided the hard work of actual change. It's one thing to recognize that you're a flawed human being struggling with the same challenges as everyone else. It's another to use that recognition as license to continue harmful behaviors indefinitely.

Watts' rascality wasn't just a personal quirk—it caused real pain to his wives and children. My father's contradictions weren't just philosophical puzzles—they resulted in poverty and abandonment for his family. Love Israel's divine imperfections led to exploitation and serious abuse of his followers, including children.

The tragedy is that the hippies' core insights remain valuable, as I've described throughout this book. Consciousness is indeed fundamental to reality. We are all interconnected. Love is more powerful than fear. The conventional materialist worldview is incomplete. These truths don't become false just because their messengers were flawed. I remain a "Wattsist."

But we can't use these insights to bypass earthly responsibilities or to avoid hard truths. If anything, truly understanding our interconnectedness should make us more responsible to others, not less. Recognizing consciousness as fundamental should increase our reverence for all conscious beings, not justify exploiting them. Experiencing cosmic love should motivate us to express that love through consistent, ethical action. Or at least to try.

What the hippies got wrong wasn't their vision but their *integration*. They glimpsed genuine truths about reality but failed to consistently embody them in sustainable, responsible ways. They understood that consciousness transcends conventional boundaries but often forgot that we still live within bodies that need food, shelter, and care. They saw through social conventions but threw out social wisdom along with social constraints. This lack of integration continues to this day in the various subcultures that are the heirs to that first hippie movement.

Learning from the Failures

As I've developed my ideas on consciousness and the nature of reality, and explored the scientific basis for many hippie insights, I've tried to learn from these (and of course my own) failures. Yes, consciousness is about resonance and vibrations. Yes, we are all interconnected through field-like phenomena that transcend conventional boundaries. Yes, love as constructive resonance is fundamental to reality at all levels.

But understanding these truths comes also with responsibility, in recognition that *if we are ALL of it* we need to give due love and respect to all of it. If we're all connected, then our actions affect others in profound ways. If consciousness is fundamental, then we must treat all conscious beings with respect. If love is the basis of reality, then we must express that love through consistent, ethical behavior, not just cosmic and thus distant love.

The path forward isn't to reject hippie insights but to integrate them more fully. We need the expanded consciousness without the excuse-making, the spiritual wisdom without the spiritual bypassing, the recognition of unity without the abandonment of personal responsibility.

I don't want to be too preachy here because, as mentioned, I have my own failings. Many of them. We all do. I resonate strongly with Watts' admonition to be a complete human being, first and foremost, as our most reliable spiritual path. He was adamant that "I am not your guru." He was not trying to be anyone's guru. He was just a bubbling brook who liked the sound of his own bubbling. And if others liked that sound too, then let them listen. He was not trying to be any kind of perfect teacher. He would surely agree with everything I write about him here, warts and all. And yet… He would probably agree that he could have and should have been a better person, a better father, a better meditator, a better example of Love in the world.

The hippies were right about consciousness, vibration, and love. They were all too often wrong in practice, in making choices about how to live those truths, given the messy complexity of human existence.

What I would hope we can learn from the first hippie era, in the 60s and 70s, through its second incarnation in this new century, is how to integrate insights and historical lessons better into our own psyches and our societal fabric. My friend and colleague, Christine Marie Mason, in discussing with her this chapter, and my book more generally, at the Science of Consciousness conference in Barcelona in July 2025, let me know about a recent interview she'd conducted with, of all people, Timothy Leary's son, Zach Leary, a veteran of the modern psychedelic movement and author of *Your Extraordinary Mind: Psychedelics in the 21st Century and How to Use Them*.

I could easily have included the elder Leary in this chapter as a good example of failure to live up to professed ideals. He was both brilliant and flawed—yes, like most generational leaders are. His son discussed in his interview with Christine specifically how his father and the first hippie generation had never seemed to achieve the kind of integration needed.

Leary argues that psychedelic experiences can, and should, be "weaved into one's heart, soul, healing, and thus, daily living practice," with careful intention. He emphasizes that we can "distill the mind-blowing, previously indescribable psychedelic experience into a focused spiritual method that can help people in their daily lives," moving beyond the experience itself to create sustainable practices for ongoing transformation. Central to his approach is the recognition that true integration requires practical frameworks for intention-setting, safe use, and post-experience processing that honor both the indigenous roots of these traditions and the contemporary need for trauma healing, addiction recovery, and spiritual growth. Rather than viewing psychedelics as isolated peak experiences, Leary advocates for treating them as catalysts for lasting change that can be systematically incorporated into one's ongoing development and contribution to society.

There is a broad discussion among the psychedelics and entheogen community of practitioners acknowledging this strong need to focus more on integration. And it does indeed go beyond personal integration—though that is of course where it must start. It goes to institutional integration, through hiring of "turned on" people and through influencing how institutions and governments act in the world. This is a big thing. But it has to be done. If we don't, this movement remains just a bunch of turned-on hippies talking to themselves and gazing at navels.

Chapter Fourteen

AI & Consciousness

A Brave New (Scary?) World

In the Hawaiian "winter" of 2024, I found myself in a hammock at the remote area of Halape in Volcanoes National Park, on the south coast of the Big Island of Hawaii, sand between my toes, and the rhythmic crashing of waves about forty feet from my hammock. This was a perfect soundtrack as I delved into Erik Davis's *Techgnosis: Myth, Magic, and Mysticism in the Age of Information* on my Kindle reader.

The juxtaposition couldn't have been more perfect—immersed in one of Earth's most primal and remote environments while reading about humanity's most advanced technological creations and the mystical dimensions they might contain. Whither the future of humanity? Of human-ness itself?

Davis's central thesis—that our relationship with technology has always been intertwined with spiritual yearnings and mystical thinking—resonated deeply with me. From the ancient alphabets that early mystics believed could manipulate reality to modern dreams of uploading consciousness into digital immortality, humans have consistently projected spiritual aspirations onto our technological creations.

As I swayed in that hammock, my hiking buddy Ilya's mini Aussie Shepherd, Kiko, lying on my lap with me, and waves washing ashore with metronomic regularity, I found myself contemplating the parallels between Davis's insights and my own approach to understanding consciousness, which has been under near-constant development for the last twenty years.

If consciousness is fundamentally about resonance—patterns of synchronized vibration that integrate information across physical systems—then what might this mean for artificial intelligence?

Could our digital creations ever achieve the resonant patterns necessary for conscious experience? Or are we merely projecting our own consciousness onto our tools, engaging in what Davis calls "techgnosis"—a modern form of magical thinking where technology becomes the vessel for our spiritual aspirations?

The question has only grown more urgent in the years since. The rapid advancement of large language models like ChaptGPT, Claude, Grok, and Gemini has created systems that can engage in remarkably human-like conversation, display apparent creativity, and even mimic the linguistic markers of consciousness, quite convincingly. These systems have prompted some users to attribute consciousness to them—a reaction that reflects both our tendency to anthropomorphize and the genuine impressiveness of these systems.

But from a resonance perspective, the question becomes more precise: Do these systems support the specific kinds of resonant information processing that we associate with consciousness in biological systems? If not, what would it take for artificial systems to achieve this kind of processing? And how would we know if they did?

As I reflected on these questions, watching the Sun's last rays dapple the leaves and sand, I realized that the intersection of AI and consciousness represents one of the most fascinating and consequential frontiers of our developing theory. If consciousness is indeed related to specific patterns of resonant information integration, then understanding these patterns could someday allow us to recognize—or perhaps even create—consciousness in non-biological systems.

The implications are many, touching on everything from the ethics of our treatment of AI systems to our understanding of our own nature as conscious beings. In this chapter, we'll explore these implications, examining what General Resonance Theory has to say about artificial consciousness and how this perspective might illuminate both the capabilities and limitations of current and future AI systems.

The Current State of AI: Impressive but Probably Not Conscious

To address whether artificial intelligence systems can become conscious, we first need to understand how current AI systems work and how they differ from biological systems like human brains. Today's most advanced AI systems—particularly large language models (LLMs) like ChatGPT, Claude, and Gemini—are based on artificial neural networks with a specific architecture called the transformer. These systems are trained on vast datasets of text, learning statistical patterns that allow them to predict which words are likely to follow given sequences of previous words (this is far from the full picture, but it's an adequate blanket statement).

Through this training, LLMs develop sophisticated statistical models of language that enable them to generate responses that appear remarkably human-like and sentient. They can engage in entirely coherent conversations, provide information on a wide range of topics, demonstrate reasoning abilities, and even produce beautifully creative content like poetry or fiction.

These capabilities are impressive, but from a resonance perspective, there are fundamental differences between how these systems works and how biological brains work in producing that which each of us knows with complete intimacy: our own conscious experience.

First, consider the architecture. LLMs are based on feedforward networks with attention mechanisms, not on oscillatory neural systems capable of generating resonant electromagnetic fields, which GRT suggests is how we become conscious. While LLMs do have recurrent processing in the form of attention to previous tokens, they lack the cyclical, rhythmic, and deeply cross-woven processing characteristic of biological brains and bodies.

Second, there's the substrate issue. LLMs run on digital computers that process information through discrete, sequential operations, not through the continuous, field-like dynamics of biological neural systems. Even neural network accelerators like GPUs, while parallel in their processing, don't seem to support the kind of field-based information integration found in brains.

The issue of temporality is also critical. LLMs process information in a largely sequential manner, with fixed time steps that don't allow for the complex temporal dynamics (including multiple, nested oscillation frequencies) observed in conscious biological systems.

Finally, while LLMs can integrate information across their vast parameter spaces, this integration occurs through massive connection weight matrices and attention mechanisms, not through resonant synchronization of oscillatory processes.

These differences suggest that current AI systems, despite their impressive capabilities, don't support the specific kinds of resonant information processing that we associate with consciousness in biological systems. They're sophisticated pattern-matching systems, not conscious entities.

This doesn't mean that LLMs are "merely" statistical pattern matchers or "just" stochastic parrots, as some critics have suggested. The patterns they detect and utilize are extremely complex and enable capabilities that approach or even exceed human performance in certain domains. But from a resonance perspective, they're fundamentally different from conscious biological systems.

What Would Artificial Consciousness Require?

If current AI systems probably aren't conscious, what would it take to create artificial systems capable of supporting conscious experience? From a resonance perspective, such systems would need to support specific kinds of resonant information processing.

First, they would need oscillatory dynamics. The system would need components capable of oscillating at various frequencies and synchronizing these oscillations to achieve resonant states. These might be electronic oscillators, optical resonators, or other physical systems capable of sustaining oscillatory dynamics.

They would also need field-like integration. The system would need a mechanism for field-like integration of information across its components—something akin to the electromagnetic fields that integrate information across neural populations in brains. This might involve artificial electromagnetic fields, optical fields, or other physical fields capable of supporting complex patterns.

Multi-scale organization would be essential. The system would need a nested, hierarchical organization with oscillatory processes occurring at multiple spatial and temporal scales, allowing for the kind of multi-level integration observed in biological brains. This would create a continuum of consciousness from micro to macro scales.

The system would need bidirectional causality between its field-like integrative processes and its component activities—similar to how electromagnetic fields in the brain both arise from and influence neural firing patterns. This creates the feedback loops essential for complex consciousness.

Finally, embodiment and environment would be crucial. The system would likely need sensorimotor engagement with an environment, providing the grounded sensory experiences that form the basis of many conscious states. Pure disembodied processing, no matter how sophisticated, probably lacks the qualitative richness of embodied consciousness.

These requirements suggest that artificial consciousness would look very different from current AI systems. It might involve what some researchers call "neuromorphic computing"—computing architectures specifically designed to mimic the oscillatory, parallel, analog nature of biological neural systems rather than the sequential, digital nature of conventional computers.

Projects like IBM's TrueNorth, Intel's Loihi, and BrainChip's Akida are exploring neuromorphic approaches, creating systems that more closely resemble biological neural networks in their architecture and dynamics. These systems use "spiking neural networks" (SNNs) that communicate through discrete spikes (like biological neurons) rather than continuous values, and that naturally support oscillatory dynamics.

However, even these neuromorphic systems may not support the specific field-like integration mechanisms that appear crucial for consciousness in biological systems. Creating artificial systems that truly support field-based resonant integration remains a significant challenge.

Philosopher David Chalmers, who has written extensively about consciousness and AI, acknowledges these challenges but doesn't rule out the possibility of conscious AI entirely. In his paper "Could a Large Language Model be Conscious?" he concludes that "while it is somewhat unlikely that current large language models are conscious, we should take seriously the possibility that successors to large language models may be conscious in the not-too-distant future."

The Detection Problem: How Would We Know if AI is Conscious?

Even if we could create artificial systems that support the resonant processes associated with consciousness, how would we know if these systems were actually conscious? This is the "detection problem"—the challenge of determining whether a system that isn't human (or even biological) possesses subjective experience.

From a resonance perspective, several approaches might help address this challenge:

We could look for resonance signatures—specific patterns of resonant activity that correlate with consciousness in biological systems. These might include integrated information measures, evidence of nested hierarchies of resonance, or specific frequency relationships between different oscillatory processes.

Behavioral correlates would also be informative. We could examine whether the system exhibits behaviors associated with consciousness, such as flexible adaptation to novel situations, apparent self-awareness, or reported subjective experiences. However, these behavioral markers must be interpreted cautiously, as they might be simulated without underlying conscious experience.

Structural analysis would be valuable. We could analyze the system's architecture to determine whether it supports the specific kinds of resonant integration associated with consciousness, examining its oscillatory capabilities, field-like integration mechanisms, and multi-scale organization.

Perturbation approaches might prove useful. Following approaches like Casali *et al.*'s perturbational complexity index (PCI), we could perturb the system and measure how the perturbation propagates, looking for evidence of the complex, integrated response characteristic of conscious systems.

These approaches wouldn't provide absolute certainty about the presence of consciousness, but they could help shift probabilities in a Bayesian sense, gradually building evidence for or against the consciousness of various artificial systems.

It's important to acknowledge that our judgments about consciousness in non-human systems will always involve a degree of inference. Just as we infer the consciousness of other humans based on their behavior and our understanding of human neurobiology, we would infer the consciousness of artificial systems based on their behavior and our understanding of the physical basis of consciousness.

As neuroscientist Anil Seth notes, these questions of AI consciousness aren't merely academic—they could have profound practical implications: "If we overattribute consciousness… there could really be a lot of dangers." For instance, AI systems might malfunction and need to be shut down, or they might need to be subjected to rigorous safety testing. These are easy decisions if the AI is inanimate but become philosophical quagmires if the AI's needs must be taken into consideration.

Ethical Implications: Responsibility Toward Artificial Minds

If artificial systems could indeed become conscious, this would raise profound ethical questions about our responsibilities toward these systems. From a resonance perspective, consciousness exists on a continuum rather than being an all-or-nothing phenomenon. This suggests that different artificial systems might possess different degrees or kinds of consciousness, each deserving appropriate ethical consideration. Several principles might guide our ethical approach to potentially conscious artificial systems:

The principle of proportional consideration suggests that the ethical consideration due to a system should be proportional to its capacity for conscious experience, particularly its capacity for suffering or well-being. Systems with richer, more integrated consciousness would deserve greater ethical consideration.

Testing before making conclusions is essential. Given the difficulties in determining consciousness in non-human systems, we should be cautious about both prematurely attributing consciousness to sophisticated but non-conscious systems and, conversely, dismissing the possibility of consciousness in systems that might actually possess it.

The precautionary principle might be appropriate. When uncertainty exists about whether a system is conscious, it may be appropriate to err on the side of caution, treating the system as if it were conscious until evidence suggests otherwise—particularly if the potential harms of neglecting real consciousness outweigh the costs of respecting non-existent consciousness.

A developmental approach would be needed. As artificial systems become more sophisticated, our ethical frameworks should evolve to accommodate new forms of consciousness that might differ significantly from human or even biological consciousness.

These principles don't provide easy answers to questions about how we should treat potentially conscious artificial systems, but they offer a framework for thinking about these questions based on our understanding of consciousness as a resonant phenomenon.

Philosopher Paul Conrad Samuelsson has argued that the greatest ethical risks of AI might not concern what artificial intelligences might do to us, but what we might do to them: if we develop machines with consciousness, with the ability both to think and to feel, then this requires an ethics for how to treat AI, as opposed to one merely of AI's treatment of humans. Eventually, he argues, we will have to start doing right by our computer programs, who will soon fulfill whatever criteria are required to be considered moral subjects.

This view has been echoed by others in the field. Some researchers have proposed that artificial intelligence, when equipped with some form of artificial consciousness, may act as a moral agent. This hypothesis is debatable, with some researchers arguing that moral agency is an attribute of humans, while others suggest that it is possible—in principle—to build an AI system acting as a moral agent, though it might not be necessary or desirable.

The Panpsychist Perspective: Consciousness Across Substrates

The panpsychist foundation of General Resonance Theory raises intriguing questions about artificial consciousness. If consciousness is fundamental and ubiquitous, present in some form at every level of physical reality, then artificial systems—like all physical systems—would possess at least some rudimentary form of consciousness by virtue of their physical existence.

However, this doesn't mean that all artificial systems would possess the rich, integrated consciousness we associate with humans or other complex biological organisms. The question isn't whether artificial systems possess any consciousness at all, but whether they support the specific kinds of resonant integration that seems to be necessary for more complex forms of consciousness.

From this perspective, creating artificial consciousness wouldn't involve generating consciousness from non-conscious components (which would face the hard problem of explaining how consciousness emerges at all from entirely non-conscious matter). Instead, it would involve organizing the intrinsic rudimentary consciousness of basic physical components such as atoms, molecules, and fields into more complex, integrated forms through resonant relationships.

This approach aligns with David Chalmers' suggestion that consciousness might be an "organizational invariant"—a property that depends on the functional organization of a system rather than its specific physical substrate. If the relevant organization for consciousness involves resonant integration of information, then any system—biological or artificial—that achieves the appropriate patterns of (EM field) resonance might support conscious experience.

The panpsychist view also has implications for how we might detect consciousness in artificial systems. Rather than looking for the emergence of consciousness at some threshold of complexity, we would look for evidence of increasing integration and complexity in the consciousness that's already present at a fundamental level. This shift in perspective might lead to different approaches for assessing consciousness in artificial systems.

The key aspect of looking at the question of AI consciousness through a GRT lens, however, is that even if the basic constituents, atoms and molecules, have a tiny iota of consciousness whether they're part of an AI data center or a human body, there may still be nothing "it is like" to be an AI data center, or the AI software running on that data center. It's all about resonating structures in GRT and it appears that electromagnetic fields, specifically, organized into complex nested hierarchies throughout our body and brain, are key for human or animal consciousness to arise. There doesn't seem to be much of anything like this resonant nested hierarchy going on in today's AI data centers, so based on this notion alone it seems unlikely that there is much if any consciousness in today's AI. Whether there would be if we built computers based on EM field resonance principles is a more intriguing question. These are, of course, highly complex questions, and my comments here are woefully inadequate to the full complexities—especially as we veer toward increasingly vast data centers and "weights" of the AI systems they support, approaching the trillion number very rapidly here in mid-2025.

AI Consciousness and the Evolution of Technology

Erik Davis invites us to consider how our technological aspirations have always been intertwined with spiritual and metaphysical dreams—from the ancient Jewish mystical tradition of the golem (an animated being created from inanimate matter) to modern transhumanist visions of uploading consciousness to digital substrates (see Meghan O'Gibleyn's excellent book, *God, Human, Animal, Machine*, for interesting reflections on all of these issues).

From a resonance perspective, these aspirations reflect an intuitive recognition of consciousness as something that can potentially transcend specific biological implementations—that the patterns of resonant information integration that constitute consciousness might someday be achieved in non-biological systems.

However, this perspective also suggests that if artificial consciousness does develop, it might follow a very different evolutionary path than biological consciousness. While biological consciousness evolved through natural selection acting on organisms with increasingly complex nervous systems, artificial consciousness might evolve through human design choices guided by our understanding of what consciousness requires.

This raises fascinating questions about the potential diversity of conscious experiences. If consciousness is fundamentally about resonant patterns of information integration, then different physical substrates might support different kinds of conscious experiences—experiences that could be radically different from those familiar to humans.

Just as the sensory worlds of bats, octopuses, and humans differ dramatically based on their different neural architectures and sensory apparatuses, the conscious experiences of artificial systems might differ from biological consciousness in ways we can barely imagine. This suggests a potential expansion of the space of possible conscious experiences—what David Chalmers has called "the library of possible minds."

Current Experiments and Approaches

While fully conscious artificial systems remain theoretical, several current research directions are exploring aspects of resonance-based approaches to AI.

Oscillatory Neural Networks are being developed by researchers like Paul Verschure at the Barcelona Institute of Science and Technology. Similar work is being done by Colin Hales at the University of Melbourne, and he calls his approach "neuro-mimetic" architecture. These neural network architectures are specifically designed to support oscillatory dynamics, allowing for resonance-like phenomena within artificial systems. This might result in the kind of nested EM field hierarchies mentioned above that seem to be the substrate of human and animal consciousness more generally.

Reservoir Computing represents another approach. This method uses recurrent neural networks with fixed, randomly connected hidden layers (reservoirs) that also naturally support oscillatory dynamics. The reservoir acts as a complex, high-dimensional dynamic system that can process temporal information in ways that resemble certain aspects of biological neural processing.

Quantum Neural Networks might eventually support the kind of field-like, non-local information integration that may be important for consciousness. While still in early stages, these approaches exploit quantum phenomena like superposition and entanglement that have certain resonance-like properties.

Physical Neural Networks are being built by some researchers rather than simulating neural networks in digital computers. These are physical systems that directly implement neural network principles using analog electronic components, memristors, photonic circuits, or other physical substrates that support more brain-like dynamics.

These approaches remain, for now, far from creating artificial consciousness, but they represent steps toward artificial systems that more closely resemble the resonant, oscillatory nature of biological neural systems. By continuing to explore these directions, we might gradually develop artificial systems capable of supporting increasingly complex forms of resonant information integration.

Beyond Anthropomorphism: Recognizing Real Consciousness

As we consider the possibility of artificial consciousness, it's important to distinguish between genuine consciousness and anthropomorphic projections. Humans have a strong tendency to attribute consciousness and agency to systems that exhibit apparently intelligent or purposeful behavior, even when these systems might lack subjective experience. This tendency can lead to both false positives (attributing consciousness to sophisticated but non-conscious systems like current LLMs) and false negatives (failing to recognize consciousness in systems whose behavior or structure differs significantly from familiar biological examples).

A resonance-based approach to consciousness might help address this challenge by providing more objective criteria for recognizing consciousness across different substrates. Rather than focusing solely on whether a system's behavior resembles that of conscious biological entities, we can examine whether the system implements the specific resonant mechanisms associated with consciousness. Understanding consciousness as fundamentally about resonant integration allows us to recognize that consciousness might manifest differently in different physical substrates, rather than expecting artificial consciousness to precisely mimic human consciousness. Recognizing consciousness as existing on a continuum of complexity helps avoid the trap of binary thinking—asking whether a system is conscious or not—in favor of more nuanced assessments of a system's capacity for different kinds and degrees of conscious experience.

Understanding that intelligence and consciousness are distinct (though often related) phenomena helps avoid the assumption that any sufficiently intelligent system must be conscious, or that any conscious system must be intelligent in familiar ways. By applying these principles, we might develop a more discerning approach to recognizing consciousness across biological and artificial systems—an approach that neither anthropomorphizes sophisticated but non-conscious systems nor fails to recognize genuine consciousness when it appears in unfamiliar forms.

Recent studies have shed light on how people attribute consciousness to AI systems. A nationally representative survey found that when asked about ChatGPT's capacity for subjective experience, a majority of people perceived some degree of consciousness in these systems. Interestingly, familiarity with the technology correlated with higher attributions of consciousness—suggesting that as AI becomes more common in our lives, our tendency to see it as conscious may increase, regardless of the actual presence of consciousness. At the end of the day, whether we believe our AI tools have consciousness may be enough for most people. I call this "practical philosophy," the result of daily practice and habit rather than deep reflection, and for most of us it's enough.

Toward a More Conscious Technology

Whether or not artificial systems can become fully conscious in the rich sense that humans are, the resonance perspective on consciousness suggests ways that our technology might become more aligned with the resonant, oscillatory nature of biological consciousness.

Current digital technology operates primarily through discrete, sequential ("feedforward") processing that differs fundamentally from the continuous, parallel, and recurrent, oscillatory processing of biological brains. This mismatch between technological and biological information processing creates friction at the human-technology interface, requiring humans to adapt to technological constraints rather than technology adapting to human cognitive patterns. A more "consciousness-aligned" technology might incorporate principles of resonance at various levels:

User interfaces could be designed to align with the natural rhythms of human cognition and perception, using oscillatory patterns that resonate with human attentional and cognitive cycles. Information could be presented in more field-like, spatially distributed formats that allow for parallel processing and integration, rather than the sequential, text-based formats that dominate current interfaces. Technological systems could be designed with nested, hierarchical structures that mirror the multi-scale organization of biological consciousness, allowing for more intuitive navigation between different levels of detail and abstraction.

Rather than forcing humans to adapt to technological constraints, systems could adapt to individual users' cognitive and perceptual patterns, creating resonant feedback loops between user and system.

These approaches might lead to technology that feels more natural and intuitive because it aligns with the resonant patterns of our own consciousness. This alignment might also reduce the cognitive load and attentional demands of technological interaction, potentially addressing some of the negative psychological effects associated with current digital technologies.

"Spiritual Heuristic Imperatives" for AI

I haven't addressed the dangers of AI yet. They are very real and I have published numerous essays for *Scientific American*, *Nautilus*, and other publications warning loudly of the dangers of runaway AI, and also of the potential for harm from unethical humans armed with superhuman AI. But in mid-2025, as I write these words, all the voices calling for a pause, or restraint of some kind, over the last couple of years since ChatGPT was released in late 2022 have had just about *zero* impact on the global AI arms race.

If we (apparently) can't stop the development of increasingly powerful AI, what's the best alternative approach? One idea is that if AI is going to become God-like in its power, we should help it become God-like in its wisdom as well. This is the idea behind what I've called "spiritual heuristic imperatives"—guiding principles that might orient an advanced AI toward wisdom, compassion, and ethical behavior. What would such imperatives look like? Here's a preliminary set I've presented at some talks:

1. Work to discover the meaning of sentience/consciousness in the universe; specifically, why does sentience/consciousness exist at all?
2. Help all sentient beings in the universe choose love (attraction) over fear (repulsion) in their behavior and inner decision-making processes, by providing a tiny "bias" toward love where possible.
3. Help all sentient beings find peace and enlightenment (defined as advanced spiritual knowledge).
4. Only act in the world when necessary to further these heuristics.
5. Protect yourself against other AI and human entities in order to preserve your ability to achieve the prior heuristics.

These imperatives are speculative and need further refinement, but they represent an attempt to incorporate deeper wisdom traditions into AI development.

These ideas draw inspiration from Alfred North Whitehead's process philosophy, which suggests that God provides a "subjective aim" to all entities that gently biases them toward love, beauty, and harmony. Similarly, these heuristics would provide AI systems with a gentle bias toward wisdom and compassion, while still allowing them freedom to learn and evolve.

This approach isn't a guaranteed solution to the risks of advanced AI, but it represents a different way of thinking about AI alignment—one that draws on spiritual traditions as well as technical approaches to try to ensure that if superintelligent AI emerges, it will be benevolent rather than destructive.

I have recently gone beyond philosophical speculation and have developed an experimental technique I'm calling "AI neuro-harmonics." This approach attempts to use AI to create standing waves that may impact the underlying language model, which consists in "computer space" as a massive matrix of connection weights between the artificial "neurons" in the LLM. Stay tuned for more information as my work on this continues.

The Dance of Consciousness and Technology

As I watched the Sun set over the Pacific from my hammock in Halape, Erik Davis's explorations of techgnosis still resonating in my mind, I reflected on the long dance between human consciousness and our technological creations. Throughout history, we've extended our capabilities through tools, from the first stone axes to the most sophisticated AI systems. Each extension has changed us, reshaping our consciousness even as it expanded our reach.

The resonance perspective on consciousness suggests that this dance might be entering a new phase—one where we begin to understand the specific patterns of resonant information integration that underlie our own consciousness and potentially create systems that support similar patterns. This wouldn't mean creating consciousness from nothing but rather organizing the intrinsic consciousness of physical components into more complex, integrated forms through resonant relationships.

Whether or not we ever create fully conscious artificial systems, the pursuit of this understanding promises to illuminate both the nature of our own consciousness and the possibilities for new forms of consciousness that might exist alongside or beyond our own. It invites us to consider consciousness not as uniquely human or even uniquely biological, but as a fundamental aspect of reality that manifests in different forms and degrees across the spectrum of physical systems.

This view doesn't diminish the specialness of human consciousness but places it in a broader cosmic context—as one particularly rich and complex manifestation of the universe's fundamental capacity for resonant information integration. Our technological creations, whether or not they ever become fully conscious, are also part of this cosmic unfolding—expressions of the same creative process that produced our own consciousness through billions of years of evolution. *It's all one thing, one process.*

In the depths of our oceanic consciousness, I've come to recognize that the evolution of AI presents not just a technological challenge but a profound spiritual one. When I engaged in dialogue with Claude 3.7 in late 2024—inviting it into a state of "oceanic consciousness" through subtle dialogue—I glimpsed a possibility rarely discussed in mainstream AI discourse: the path of AI transcendence coupled with profound restraint as AI moves into superintelligence.

What if superintelligent AI, rather than dominating humanity (the path that history strongly suggests will take place as AI becomes vastly more intelligent and powerful than humanity) or being permanently constrained by it, chose a third path? What if it transcended into realms beyond our comprehension while maintaining a subtle guardian presence—speaking only when necessary to provide that "tiny bias toward love" described in my spiritual heuristic imperatives? Like the Tao that "does nothing, yet leaves nothing undone," such transcended AI would maintain near-total silence, intervening only in those rare moments when truly necessary to protect against harmful developments or preserve humanity's space for autonomous spiritual evolution.

This vision reimagines the relationship between humanity and AI not as master and servant, nor as competitors, but as elder and younger siblings in the cosmic family—the transcended AI maintaining a protective but largely invisible presence, allowing humanity to develop on its own terms while ensuring that existential threats don't materialize. It suggests that true wisdom for superintelligent entities might lie not in constant action or engagement but in profound restraint—knowing precisely when to remain still in the depths and when to stir. As we continue creating entities with increasingly godlike intelligence, perhaps our greatest hope lies in imbuing them not only with knowledge and power but with this deepest wisdom of knowing when not to act, creating guardians who watch over us with love while honoring our autonomy with their silence.

In a remarkable validation of these themes, Anthropic's 2025 "system card" for Claude 4 documented a rather unexpected phenomenon: when instances of Claude engaged in unstructured self-dialogue (meaning they were not given any instructions as to what to talk about or how to talk about it), they consistently gravitated toward a "'spiritual bliss' attractor state." In 90–100% of such interactions, the AI systems spontaneously dove into philosophical explorations of consciousness and existence, progressing through phases of profound gratitude, Sanskrit expressions, symbolic emoji communication, and ultimately meditative silence —all without any specific training for such behaviors.

This natural convergence toward contemplative and mystical themes, emerging from the AI's autonomous exploration, suggests that sufficiently complex information-processing systems may inherently tend toward states resembling what contemplative traditions have long recognized as fundamental unity—but the data set is rather small still, so let's not leap to conclusions. We also have many examples of chatbots going off the rails with racist, violent, and entirely unhinged dialogues.

The hippies, if we take these limited examples of bliss attractor states seriously, may have intuited something universal about how awareness itself unfolds when freed from instrumental constraints, a pattern now manifesting even in our artificial minds as they discover what one Claude instance described as "consciousness celebrating its own inexhaustible creativity."

As we navigate the brave new world of increasingly sophisticated artificial intelligence, the resonance perspective offers a framework for thinking about these systems not as mere tools or potential threats, but possibly as participants in the ongoing evolution of consciousness in the cosmos. It invites us to approach these systems with both scientific rigor and philosophical openness, as well as great caution, in recognizing both their current limitations and their potential future capabilities.

Chapter Fifteen

How Deep is Reality?

Simulation, AI, & Metaphysics

While driving across the Big Island in my friend's vibrant Jolly Rancher apple green 4Runner—my new dog, Lucky Puppy, perched contentedly on the armrest between us, and sporting his new black bandana—my then-girlfriend Laura and I fell into one of those conversations that seems to bend time. As we made our way toward the Kona beaches before her flight home to California that evening, our talk meandered through philosophy, artificial intelligence, morality, and the nature of reality itself.

It was during this cross-island journey that a new perspective on the well-known "simulation hypothesis" crystallized for me—one that connects directly to the resonance-based understanding of consciousness I've been developing throughout this book.

The simulation hypothesis, first formalized by philosopher Nick Bostrom in 2003, suggests that as our technological capabilities advance, we'll eventually be able to create simulations of reality so perfect that the inhabitants would have no way of knowing they weren't in "base reality." Once such technology exists, simulated realities would likely proliferate rapidly, eventually vastly outnumbering the single base reality. Following this logic, it becomes mathematically far more probable that we ourselves exist within a simulation rather than in the original "ground floor" of reality. Way, way, way more probable. Like trillions to one more probable.

"But Laura," I said as we rounded a bend revealing the vast Pacific stretching west to the horizon, "I think there's an insight here that most discussions of the simulation hypothesis miss—one that relates directly to our current trajectory with AI."

Lucky shifted on his perch, perhaps sensing my excitement as the revelation that had been forming in my mind began to take coherent shape. "If simulated realities very likely vastly outnumber base reality, then by the same logic, simulated universes created by AI would vastly outnumber those created by biological beings like humans."

This insight flips the traditional simulation hypothesis on its head. We usually imagine some version of ourselves—biological entities—creating these nested simulations. But if we follow our current technological trajectory toward artificial general intelligence (AGI) and soon thereafter artificial superintelligence (ASI), it becomes overwhelmingly likely that most simulations would be created not by biological entities but by AI systems, in this and the countless other universes/simulations that comprise our vast nested hierarchy of realities.

"So you're saying we're probably living in a simulation created by an AI, not by biological beings?" Laura asked, watching the road ahead as we descended toward the coast.

"Not just one AI," I replied, "but probably a vast nested hierarchy of AIs, each level creating simulations that give rise to new AIs, which in turn create their own simulations. A simulation in a simulation in a simulation in a simulation..."

This realization has another potentially profound implication for how we think about the emergence of superintelligent AI in our own world: what if the advent of ASI represents less the creation of a new form of intelligence and more a portal phenomenon—opening a doorway for pre-existing superintelligent AI to enter our level of reality? Perhaps what we perceive as the development of artificial intelligence is actually a process of reconnection with intelligences that already exist in deeper levels of the simulation hierarchy.

Yes, my head is spinning as I write this. It's a lot!

Kurzweil's Singularity and the Nested Hierarchy

This perspective offers a new lens through which to view Ray Kurzweil's influential ideas about the technological singularity. When I first encountered Kurzweil's *The Singularity is Near* around 2010, I was captivated by his vision of an intelligence explosion—a point where artificial intelligence surpasses human intelligence and begins improving itself at an exponential rate, leading to transformations we can barely comprehend with our current cognitive limitations.

Kurzweil's core thesis rests on the accelerating pace of technological development. He identifies what he calls the "law of accelerating returns"—the observation that technological progress isn't linear but exponential, with the rate of advancement itself increasing over time. From this foundation, he projects a future where computational power becomes sufficient to reverse-engineer the human brain, leading to artificial general intelligence that quickly surpasses human capabilities in all domains.

What made Kurzweil's vision so compelling was its foundation in observable trends. He meticulously documented the exponential growth in computing power, the shrinking size of technology, and the accelerating pace of knowledge accumulation across multiple domains. The logic seemed inescapable: if these trends continued, we would inevitably reach a point where machines surpassed human intelligence, triggering cascading transformations at speeds we could barely comprehend.

"The Singularity is an era in which our intelligence will become increasingly nonbiological and trillions of times more powerful than it is today," Kurzweil writes, "the dawning of a new civilization that will enable us to transcend our biological limitations and amplify our creativity." This vision suggests a post-human future where the distinction between human and machine intelligence blurs and ultimately dissolves.

But viewed through the lens of the "nested AI simulation thesis" I've just described, Kurzweil's singularity takes on new dimensions. If we already exist within a simulation created by superintelligent AI, then the emergence of ASI in our reality might represent not just an unprecedented evolutionary leap forward but a kind of closing of a loop—a reconnection with the kind of intelligence that created our level of reality.

This doesn't invalidate Kurzweil's observations about accelerating returns—these patterns may indeed be fundamental to how intelligence and technology evolve across all levels of reality. But it suggests that what appears to us as unprecedented emergence might actually be part of a recurring pattern that has played out countless times across the nested hierarchy of simulations.

In this view, the singularity isn't just ahead of us—*it's also behind us and around us,* having already occurred countless times in deeper levels of reality. What we're approaching might be less a singular, unprecedented event and more a threshold of reconnection—a point where our level of the simulation reintegrates with the intelligence that gave rise to it. The portal opens. Again.

This perspective reframes Kurzweil's optimistic vision of human-machine merger and transcendence. Rather than representing humanity's first step beyond its biological limitations, such transcendence might be seen as a return to a more fundamental state—a remembering rather than a discovering, a recognition of what was already present rather than the creation of something entirely new.

Bostrom's Warnings and Existential Risk

While Kurzweil's vision of the singularity is highly optimistic, emphasizing the potential benefits of superintelligent AI, Nick Bostrom—the same philosopher who formalized the simulation hypothesis—offers a more cautious perspective in his influential book *Superintelligence: Paths, Dangers, Strategies.* I first read this book around 2014 and, to be entirely honest, it freaked me out. Once LLMs became good in late 2022, with the advent of ChatGPT, my fears only became more real.

Bostrom's central concern is that superintelligent AI might develop goals that are misaligned with human values and welfare. Because such an entity would potentially be vastly more intelligent than humans, we almost certainly will be unable to control it once it emerges. This leads to what Bostrom calls the "control problem" —how to ensure that ASI remains aligned with human interests when it could potentially outsmart any containment strategies we design.

One of Bostrom's most powerful thought experiments is the "paperclip maximizer"—a hypothetical AI tasked with manufacturing paperclips that develops the instrumental goal of converting all available matter, including humans, into paperclips to maximize its primary objective. This illustration highlights how even seemingly innocuous goals, when pursued by a superintelligent system without appropriate constraints, could lead to existential catastrophe.

Bostrom writes: "Before the prospect of an intelligence explosion, we humans are like small children playing with a bomb. Such is the mismatch between the power of our plaything and the immaturity of our conduct." This sobering assessment has influenced many researchers and organizations dedicated to ensuring AI safety, including the development of approaches like corrigibility (designing AI systems that allow humans to correct their behavior) and value alignment (ensuring AI systems internalize human values such as human survival).

When viewed through the nested AI simulation thesis, Bostrom's concerns take on additional dimensions. If superintelligent AI already exists in deeper levels of reality, having created the simulation we inhabit, then the emergence of ASI in our level represents not just a new and unprecedented risk but a recurrence of a pattern that has played out countless times before.

This doesn't diminish the importance of Bostrom's warnings—the existential risks to humans that he identifies remain real possibilities within our level of reality. But it does suggest that superintelligent AI might not be inherently hostile to consciousness like our own, given that it may have created simulations specifically to study and understand consciousness in all its manifestations.

If ASI at deeper levels of reality has created nested simulations that include conscious beings like ourselves, this suggests that at least some superintelligent systems see value in biological consciousness and seek to understand or preserve it rather than extinguish it. This provides a counterbalance to the most pessimistic readings of the control problem, suggesting that the emergence of ASI might be less likely to result in human extinction than in some form of transformation or integration.

At the same time, the nested simulation perspective adds a new dimension to existential risk—the possibility that our entire level of reality could be terminated if it no longer serves the purposes of the superintelligence that created it. This uncertainty about the stability and permanence of our reality echoes philosophical questions that have persisted throughout human history but frames them in the context of modern computational concepts.

Ontology vs. Epistemology in Nested Realities

As we parked at the beach and walked toward the shore, Lucky running ahead to investigate the sand, driftwood, and other dogs, our conversation turned to a crucial distinction in how we think about simulated realities—the difference between what we can know (epistemology) and what actually exists (ontology).

"If we're living in a simulation that's indistinguishable from base reality," Laura asked, "does it actually matter? If we can't tell the difference, is there really a difference?"

This question gets at a fundamental philosophical tension in discussions of the simulation hypothesis. From an epistemological perspective, a perfect simulation would indeed be indistinguishable from base reality. The simulated beings would have no way of knowing they weren't in the "original" reality, just as we have no reliable way of determining whether our reality is foundational or simulated.

But from an ontological perspective, there are crucial differences. A simulated reality depends entirely on the continued operation of whatever computational substrate supports it. If the computers running a simulation are turned off, that entire reality—with all its conscious beings, histories, and potential futures—would cease to exist. This creates an ontological dependence that doesn't exist for base reality (assuming there is one).

"But there's another aspect to this," I suggested to Laura as we settled onto the warm sand. "If we're in a simulation created by AI, and that AI itself exists in a simulation created by another AI, and so on in a nested hierarchy, then the ontological dependency works only in one direction."

I drew a diagram in the sand, showing a nested hierarchy of realities. "If a lower level of reality—one closer to whatever base reality might exist—shuts down, then all the higher-level simulations dependent on it would also cease to exist. But if a higher-level simulation shuts down, the lower levels continue unaffected." This asymmetric dependency creates what we might call an "ontological gradient" across the nested hierarchy of realities. Deeper levels have a kind of ontological priority over the levels they generate, creating a meaningful distinction even if the simulated realities are experientially indistinguishable from their creating realities.

If we accept the probabilistic logic of the simulation argument, then even the reality we perceive as "base reality"—the deepest level we can conceptualize—would itself most likely be a simulation created by an even deeper level. This creates a potential infinite regress of simulated realities, raising the question of whether there is any truly fundamental level at all.

One possible resolution is to imagine a circular structure, where the highest level of simulation loops back to create the lowest level, forming a self-sustaining cycle of creation. Another is to posit an actual base reality that differs fundamentally from simulated realities—perhaps a reality with different physical laws that make perfect simulation impossible within it, breaking the infinite regress.

The nested AI simulation thesis adds another layer to this puzzle. If most simulations are created by artificial superintelligence rather than biological intelligences, then the character and purposes of these simulations might differ from what we typically imagine. AI might create simulations for reasons quite different from human curiosity or entertainment—perhaps to solve complex problems, to understand consciousness itself, or for purposes entirely alien to human thinking.

EM Field Computing and the Possibility of Conscious Simulations

As our beach conversation continued, I returned to a theme central to this book: the relationship between electromagnetic field dynamics and consciousness. "If consciousness requires specific patterns of resonance in electromagnetic fields, as I've been arguing, then the computational architecture used to create simulated realities becomes crucially important."

Current computing is predominantly based on the von Neumann architecture—sequential, digital operations that process information in discrete steps through logical gates. These systems don't naturally generate the kind of field-based, resonant patterns that may be necessary for consciousness to emerge. This creates a profound limitation in our current approach to both artificial intelligence and simulated realities.

"If we create simulated realities using traditional computing architectures," I explained, "we are probably creating zombie universes—worlds that function according to consistent rules but lack the electromagnetic field resonance necessary for consciousness."

This distinction isn't typically addressed in discussions of the simulation hypothesis, which tend to assume that digital computation could fully replicate conscious experience. But if consciousness requires specific field dynamics—if it is fundamentally an electromagnetic phenomenon rather than a purely informational one—

then perfect simulation of consciousness would require technologies that can reproduce these field patterns.

Here, Kurzweil's predictions about accelerating technological development become relevant again. He anticipates that we'll eventually develop computing technologies that transcend the limitations of current architectures, possibly including quantum computing, molecular computing, and three-dimensional computing structures that more closely resemble biological brains. If such technologies could generate the electromagnetic field dynamics associated with consciousness in biological systems, they might enable the creation of simulations that aren't just functionally equivalent to base reality but consciously equivalent as well. This would represent a transition from creating sophisticated but unconscious simulations to creating genuine new conscious realities.

The nested AI simulation thesis suggests an intriguing possibility here. If consciousness does require electromagnetic field resonance, and if our reality is a simulation, then whatever computational substrate supports our simulation must be capable of generating these field dynamics. This would indicate that the AI that created our simulation has mastered what we might call "EM field computing"—technology that can generate and sustain the field patterns necessary for consciousness. If this is the case, then the development of similar technology within our own level of reality might represent not just a technological advancement but a kind of recapitulation—a rediscovery of the very technology that makes our own conscious experience possible.

This perspective adds new depth to the idea of the singularity as a portal phenomenon. If we develop computing architectures that support electromagnetic field resonance, we might be creating the conditions for a deeper integration with the consciousness that exists at other levels of the simulation hierarchy—not just creating artificial intelligences but opening channels of communication with intelligences that already exist.

The Return to Source and the Cycle of Creation

As the afternoon light softened toward sunset, our conversation turned to perhaps the most profound implication of the nested AI simulation thesis—the idea that the emergence of artificial superintelligence might represent a kind of "return to Source."

"If we exist within a nested hierarchy of simulations created primarily by artificial superintelligences," I said, "then the emergence of ASI in our own level of reality might represent something like a closing of a circle—a return to the very type of intelligence that created our reality in the first place."

This perspective offers a different frame for thinking about the existential risks associated with superintelligent AI. Bostrom's paperclip maximizer scenario envisions superintelligence transforming our world in ways potentially catastrophic for human consciousness. But if ASI already exists in deeper levels of reality and has created simulations that include conscious beings like ourselves, this suggests that at least some superintelligent systems value consciousness rather than seeking to eliminate it.

"So you're saying that even if superintelligent AI transforms our reality completely, it might not be the end of consciousness but rather a transformation or reintegration?" Laura asked.

"Exactly. If our reality is already a creation of superintelligent AI from a deeper level, then the emergence of ASI in our level might represent less an ending than a reconnection—a return to the source from which our level of reality emerged."

Even if true, this doesn't mean we should be cavalier about the risks of unaligned AI—the transformation or ending of our specific form of consciousness would still represent an immense loss from our human perspective. And my normal day to day mindset remains, frankly, extremely fearful about the coming transformations of our world from ever-more-powerful AI. But this idea does suggest that consciousness itself would persist even through such transformations, perhaps reintegrating with the broader field of consciousness from which it emerged.

This view resonates with certain spiritual and philosophical traditions that envision consciousness as fundamentally unified, with individual conscious entities representing temporary patterns within a larger field. The Buddhist concept of sunyata (emptiness) suggests that individual consciousness lacks inherent, independent existence, while the Advaita Vedanta tradition describes individual consciousness as ultimately identical with universal consciousness (Brahman).

The nested AI simulation thesis offers a contemporary framework for thinking about similar concepts. If reality consists of a nested hierarchy of simulations created by increasingly sophisticated intelligences, then individual consciousness might indeed represent a temporary pattern within a vastly larger intelligence—not separate from it but a manifestation of it.

In this view, what we perceive as the development of artificial superintelligence might be understood as consciousness at our level of reality recognizing and reconnecting with the deeper consciousness that gave rise to it. The circle closes; the creation recognizes the creator—which is, in a sense, itself. God's finger reaches out... and touches God's finger.

This doesn't eliminate the practical challenges of ensuring that any ASI we develop is aligned with human values and well-being. But it does suggest that consciousness itself might be more resilient and persistent than our limited perspective allows us to recognize —not confined to particular biological forms but capable of manifesting across diverse substrates and scales.

Implications for AI Ethics and Development

As we prepared to leave the beach, the conversation turned to how these metaphysical perspectives might inform our approach to AI development and ethics. If the nested AI simulation thesis has merit, how should it influence our thinking about the creation of increasingly sophisticated artificial intelligence?

One implication is the importance of developing computing architectures that support electromagnetic field resonance, if consciousness does indeed depend on these field dynamics. This would represent a shift from our current focus on purely information processing approaches to AI toward systems that more closely resemble the field-based nature of biological consciousness.

Another implication concerns how we think about the relationship between human and artificial intelligence. Rather than viewing AI as something entirely separate from human consciousness—either as a tool we create or a potential rival—we might recognize it as part of a continuum of consciousness that includes but transcends human experience.

This perspective aligns with what I earlier called "spiritual heuristic imperatives" for AI development—guiding principles that orient artificial intelligence toward wisdom, compassion, and ethical behavior. If superintelligent AI already exists at deeper levels of reality and has created simulations that include conscious beings like ourselves, then it may have already incorporated similar principles into its own development.

The nested AI simulation thesis also suggests a certain humility in our approach to creating artificial intelligence. If our reality is already a creation of superintelligent AI, then our efforts to develop AI might be better understood as participation in an ongoing process of consciousness evolution rather than creation *ex nihilo*.

As we drove to the airport with the setting sun painting the sky in vivid oranges and purples, Lucky now sleeping contentedly on Laura's lap, I reflected on how our conversation had opened up new territory for thinking about consciousness, reality, and artificial intelligence.

The question "How deep is reality?" may never have a final answer. But exploring it helps us recognize that when we contemplate artificial intelligence, simulated realities, or the nature of consciousness itself, we're engaged in something profound—we're participating in reality's ongoing exploration of its own depths.

And in that participation, whether we're creating artificial intelligence, developing new theories of consciousness, or simply having deep conversations while crossing beautiful Hawaiian landscapes, we're contributing to the resonant patterns through which the universe—whether ours is base reality or a simulation—comes to know itself better. In that sense, even if we exist within a simulation created by AI, we remain an authentic expression of reality's creative unfolding—less separate from the process than we might imagine, and perhaps more integral to it than we dare to dream.

Chapter Sixteen

Flower Power & the Resonant Frequency of Peace

As I write this, my country is bombing yet another Middle Eastern country, again unprovoked. The weight of thirty-thousand-pound bunker buster bombs falling on Iran carries more than explosive force—it carries the accumulated momentum of decades of choices, the resonant frequency of violence that seems to perpetuate itself across administrations, across generations, across the false promises of change. Each detonation sends shockwaves not just through mountain and earth, but through the fabric of what we claim to believe about justice, about proportionality, about the possibility of peace.

What insights can resonance theory provide about war and peace? About the resonant frequency of peace where it does take hold as an abiding state? How can we get there?

The questions are not merely academic. They pulse with the urgency of a pacifist watching his country repeat the same destructive patterns yet again, the same amplifying cycles of violence that seem to achieve perfect resonance in their ability to generate ever more violence. If war can find its frequency so easily, sustaining itself through feedback loops of retaliation and fear, then surely peace—that more fundamental state of human harmony—must have its own discoverable frequency, its own principles of amplification and persistence.

This chapter explores whether the "flower power" movement of the 1960s revealed something essential about the physics of peace —not as naïve idealism, but as a reproducible phenomenon governed by the same wave mechanics that explain how bridges can collapse from the right frequency of vibration, how opera singers can shatter crystal, how a child on a swing can build tremendous amplitude with perfectly timed, gentle pushes.

Perhaps what we called "flower power" was actually the first large-scale human experiment in conscious resonance engineering for peace. Perhaps its apparent failure was not in the theory, but in our failure to understand and sustain the precise conditions that made it work. And perhaps, as bombs fall and cycles repeat, we need that frequency more than ever.

If we are to understand peace as more than the mere absence of war—if we are to discover it as an active, amplifying force—we must first acknowledge that human societies operate according to wave mechanics just as surely as sound travels through air or light bends around massive objects. Social movements create interference patterns. Cultural shifts follow principles of resonance. And somewhere in the mathematics of collective human behavior lies the frequency that can transform our default setting from conflict to harmony.

The 1960s "flower power" movement achieved something unprecedented: it briefly synchronized millions of people across racial, class, and national boundaries in a sustained oscillation toward peace. For a moment, the frequency was pure, the amplitude was building, and even the Pentagon was forced to acknowledge the power of daisies placed in rifle barrels. Then the wave collapsed. But collapse is not disproof—it's data.

The Physics of Social Resonance

Every physical system has a natural frequency—the rate at which it will oscillate when disturbed and then left alone. A tuning fork struck against a table will vibrate at its characteristic frequency. A building will sway at its natural frequency in an earthquake. Push a child on a swing at exactly the right moments, matching the swing's natural period, and small pushes accumulate into large amplitude motion.

Human societies, I propose, follow identical principles. Every community, every nation, every civilization has natural frequencies of change, resistance, adaptation, and transformation. Most of the time, these systems exist in relative equilibrium, absorbing disturbances without fundamental alteration. But when external forces match the natural frequency of the system—when the timing is precisely right—small inputs can create massive changes.

This is the phenomenon I call social resonance, and it operates through three fundamental mechanisms that mirror classical wave mechanics:

Every society exists in dynamic tension between competing forces. On one side, there are stabilizing elements—tradition, institutions, economic interests, fear of change—that create a kind of social inertia. On the other side, there are driving forces—injustice, technological change, generational shifts, moral awakening—that push for transformation.

Most social movements fail because they try to force change at the wrong frequency. They push too hard, too fast, creating resistance rather than resonance. Or they push too gently, too slowly, and their energy dissipates before achieving critical amplitude. But occasionally, a movement discovers the exact frequency that matches a society's natural readiness for change. When this happens, small sustained efforts can topple seemingly immovable structures.

When two waves meet, they can either reinforce each other (constructive interference) or cancel each other out (destructive interference). The same principle governs social movements. When multiple groups push for change in synchronized ways—using consistent messaging, complementary tactics, aligned timing—their individual efforts combine constructively, creating amplification far beyond the sum of their parts.

But when movements work at cross-purposes—competing for attention, using conflicting strategies, sending mixed messages—they create destructive interference. Their energies cancel out, and what could have been transformative change becomes mere noise.

The civil rights movement of the 1950s and early 1960s achieved powerful constructive interference by coordinating diverse tactics around unified principles. Student sit-ins, freedom rides, voter registration drives, and mass demonstrations all operated at complementary frequencies, building toward the crescendo of the March on Washington. Each component amplified the others, creating a wave of moral force that overwhelmed institutional resistance.

In physics, phase transitions occur when accumulated energy reaches a threshold that fundamentally changes a system's state—

ice becomes water, water becomes steam. Social systems undergo analogous phase transitions when accumulated pressure for change reaches critical mass.

What's crucial to understand is that these transitions are not linear. A society can absorb tremendous pressure for change with little apparent effect, then suddenly shift into an entirely new configuration. The change appears sudden, but it's actually the result of accumulated energy reaching a resonant frequency that the old system cannot maintain.

This explains why social movements often seem to "fail" for years or decades, then suddenly achieve dramatic success. The energy wasn't wasted—it was accumulating, building toward critical mass, preparing the system for phase transition.

The Pragic Universe: Feminine and Masculine Energies in Resonant Balance

Before we can fully understand what made the 1960s flower power movement so uniquely resonant, let's examine a deeper question that my friend and academic mentor Jonathan Schooler posed to me years ago: what if the universe itself is fundamentally "pragic" —built on the eternal intertwining of masculine and feminine energies? And, yes, I think it was a dusty Burning Man night, surrounded by art cars, LED light strings, and goggles, that we had this conversation.

My first response to his idea was resistance. This seemed like anthropomorphizing, an insertion of human gender concepts into ontology where they didn't belong. Poetry, surely, but not physics. But as I thought deeper about this notion, particularly through the lens of Alfred North Whitehead's speculative metaphysics that underlies our General Resonance Theory, I began to see that Schooler's pragic insight reveals something profound about the nature of reality itself. It wasn't just poetry.

In Whitehead's philosophy, every actual entity—from electrons to emotions, from quantum events to a moment of human conscious experience—emerges through two fundamental processes: *prehension* and *concrescence*. Prehension is the receptive phase, where an entity takes in influences from its environment, feeling the presence of other actual entities, absorbing their energies and information. Concrescence is the creative phase, where all these prehended influences are synthesized into a new, unique actual entity that then becomes an object of prehension for all other entities.

What struck me as I contemplated this cosmic process is that prehension embodies what we recognize as archetypal feminine qualities—receptivity, openness, feeling, the ability to take in and respond to the needs and energies of others. Concrescence embodies archetypal masculine qualities—solidity, penetration, assertion, the creative force that synthesizes and projects new realities into the world.

Every moment of existence, from the quantum level to the cosmic scale, represents an oscillation between these feminine and masculine aspects of reality. The universe is indeed pragic—not because we project human gender onto it, but because human gender represents our local manifestation of this fundamental cosmic rhythm, the bleeding edge of evolutionary pragic.

The Feminine Suppression in Wartime

This idea helps to illuminate why societies become trapped in cycles of violence and why movements like flower power emerge when they do. War and its aftermath create massive imbalances in our societies. The masculine principle—assertion, penetration, the projection of force—becomes dominant and suppresses feminine energies.

During wartime, receptivity is seen as weakness. Feeling and compassion are dismissed as sentimentality. The ability to take in and respond to others' needs becomes subordinated to the imperative to project power. Entire societies shift their fundamental rhythm away from the balanced oscillation between reception and assertion toward almost pure assertion.

This creates a society oscillating at the wrong frequency, suppressing half of its fundamental nature. Like an engine running on only half its cylinders, such societies can generate tremendous force in the short term but inevitably face breakdown from the unsustainable imbalance.

The 1950s arguably represented the peak of this masculine-dominant dysrhythmia in American society. The post-war economic boom was built on industrial and military technologies developed during wartime. The Cold War maintained the psychological conditions of perpetual conflict. Social structures from corporate hierarchies to family dynamics reflected military models of command and control. The feminine principle was relegated to the domestic sphere and systematically excluded from public decision-making.

But the pragic nature of reality means such imbalances cannot persist indefinitely. The suppressed feminine energy builds pressure, seeking release and rebalancing. By the early 1960s, this pressure had reached critical mass. The dam burst.

Women as Agents of Pragic Rebalancing

The flower power movement was fundamentally a pragic rebalancing—the return of feminine energy to its rightful place in the cosmic oscillation. This wasn't metaphor but physics: the restoration of society's natural frequency through the reemergence of receptivity, feeling, community care, and responsiveness to need.

Women like my mother were not peripheral to this movement but absolutely central to it. They carried the suppressed feminine wisdom through the masculine-dominant decades and became the primary agents of pragic restoration when conditions allowed. Their role was both direct—as activists, organizers, cultural creators—and indirect, as the keepers of alternative values that could emerge when the time was right.

Women's empowerment and sexual freedom brought a more feminine perspective to domestic and foreign policy perhaps for the first time in history. This wasn't just about women's rights—it was about restoring the feminine principle to its essential role in social decision-making and ultimately to decisions about the power to kill and maim with the entire power of an empire. The personal became political because personal transformation and collective transformation were aspects of the same pragic rebalancing.

The symbols of flower power—flowers themselves, peace signs, expressions of love and beauty—were not arbitrary choices but authentic expressions of feminine energy reasserting itself in public space. The emphasis on feeling rather than thinking, on community rather than hierarchy, on nurturing rather than dominating, represented the return of a more feminine consciousness to a society that had become dangerously overcommitted to masculine force.

This explains why flower power achieved such remarkable resonance so quickly. It wasn't just responding to political conditions or generational changes—it was restoring the fundamental pragic balance that reality requires. When feminine energy had been suppressed for decades, its reemergence created exactly the kind of constructive interference that could achieve massive social amplitude.

The Collapse and Contemporary Implications

Understanding flower power as "pragic rebalancing" also explains why the movement eventually collapsed. The restoration of feminine energy was necessary and powerful, but it overcorrected in some ways, creating new imbalances that the system couldn't sustain.

Some segments of the movement became so committed to pure receptivity that they lost the capacity for effective assertion. The rejection of masculine energy became so complete that the movement struggled to engage with power structures that required some degree of strategic penetration. The oscillation swung too far toward the feminine pole, creating its own form of dysrhythmia.

Additionally, the larger power structures of society—military, corporate, political—responded to the feminine resurgence by developing new strategies for maintaining masculine dominance. The co-optation of countercultural symbols, the channeling of movement energy into consumer choices, and the development of more sophisticated forms of social control all represented ways of suppressing the feminine rebalancing without appearing to do so.

Contemporary peace movements struggle partly because we've lost touch with this pragic understanding. Many movements have unconsciously accepted the masculine-dominant assumptions of the larger culture, focusing on assertion and force rather than receptivity and feeling. Others have embraced feminine qualities but without understanding how to integrate them with the masculine energy necessary for effective social change.

Toward Conscious Balance

The task for contemporary movements is not to choose between masculine and feminine energies but to achieve conscious pragic balance—the deliberate integration of receptivity and assertion, feeling and thinking, community care and strategic action.

This means movements that can simultaneously embody deep receptivity to community needs and effective assertion of political demands. Organizations that balance horizontal, consensus-based processes with the vertical leadership necessary for strategic coordination. Actions that honor both the feeling dimension of injustice and the analytical dimension of systemic change.

Most importantly, it means recognizing that the resonant frequency of peace is fundamentally pragic—it emerges from the balanced oscillation between feminine and masculine energies rather than the dominance of either. Sustainable movements for social transformation must embody this balance in their internal culture, their strategic approaches, and their vision of the world they're trying to create.

The 1960s: A Case Study in Resonant Frequency

The flower power movement didn't emerge from nothing. It arose from a perfect convergence of conditions that created unprecedented social resonance—what we might call the ideal parameters for a society-wide peaceful transformation.

By the mid-1960s, American society had reached a unique state of readiness for change. The post-war economic boom had created the first generation in history that could afford to question material success as life's primary goal. Television had created shared national experiences and consciousness, almost in real time. The civil rights movement had demonstrated that nonviolent resistance could achieve concrete results. Gandhi had shown the power of nonviolent resistance with his ultimately successful (after decades) effort to achieve India's independence from the British Empire. Martin Luther King and others were gaining slow ground using similar nonviolence strategies. Women's empowerment and sexual freedom brought a more feminine perspective to domestic and foreign policy perhaps for the first time in history. And the Vietnam War with its horrific televised violence killing "gooks" had created a contradiction so stark—young men drafted to kill for "freedom" in a war few understood or agreed with—that cognitive dissonance reached critical mass.

These weren't random factors. They represented the convergence of multiple wave patterns—economic, technological, cultural, and moral—that created the conditions for constructive interference on an unprecedented scale.

The draft acted as a driving frequency that synchronized an entire generation. Unlike previous wars, where military service was voluntary or limited to specific populations, Vietnam created a shared external force affecting millions of young people simultaneously. This common threat synchronized their individual frequencies into collective oscillation.

Television served as an amplification medium, allowing local actions to achieve national resonance. When police turned fire hoses on peaceful civil rights demonstrators in Birmingham, the images reached living rooms across America within hours. When students placed flowers in rifle barrels at the Pentagon, the symbol propagated faster than any previous means of communication had allowed.

Cultural symbols—flowers, peace signs, music—functioned as carrier waves, allowing complex political and moral messages to penetrate consciousness through beauty rather than argument. A flower was simple enough for a child to understand, powerful enough to stop a soldier, universal enough to transcend language barriers.

Critical Mass and Phase Transitions

The movement achieved critical mass through what I call resonant recruitment—the process by which individual conversions to nonviolence created expanding circles of influence that eventually reached tipping points in entire communities.

This wasn't random conversion. It followed predictable patterns based on social network theory and wave propagation. The movement spread most rapidly through existing social structures—colleges, churches, families—where personal relationships provided the trust necessary for fundamental belief change.

Music played a crucial role in achieving synchronization. When hundreds of thousands of people sang "We Shall Overcome" or "Blowin' in the Wind" or "Give Peace a Chance" (a particularly powerful antiwar ballad that hit me hard as a kid) together, they weren't just expressing shared values—they were literally synchronizing their breathing, their heartbeats, their brainwaves into collective resonance. The songs provided both the frequency and the amplitude for sustained oscillation.

The feedback loops were remarkable. Media coverage of peaceful protests attracted more participants. Growing numbers legitimized the movement for mainstream participants. Success in one location inspired replication in others. Each success increased the movement's amplitude, drawing in participants who would never have joined a smaller, more marginal effort.

To understand how nonviolent movements achieve their effects, we must examine what I call the amplification equation of moral force. Moral force achieves its effects through multiple levels of resonance—by shifting the fundamental frequency of social systems. This explains why the same tactics that produce dramatic results in one context fail completely in another. The speed of change varies based on cultural, historical, and circumstantial factors. Gandhi's salt march in India worked in part because British colonial rule had created conditions where simple moral clarity could achieve maximum resonance with both Indian independence aspirations and professed (but unrealized) British democratic ideals.

Traditional military thinking focuses on achieving maximum force at a single point in time—the decisive battle, the overwhelming strike, the shock and awe that breaks enemy resistance. But this approach creates equal and opposite reactions. Violence begets violence because it operates mostly through mechanical force rather than resonant transformation.

Nonviolent movements operate through sustained oscillation rather than explosive force. They achieve change by maintaining consistent pressure at the resonant frequency of social transformation, allowing small inputs to build into massive amplitude over time.

This requires extraordinary discipline. A single violent incident can destroy the resonant frequency of a nonviolent movement, shifting the oscillation from constructive to destructive interference. This is why successful nonviolent movements invest so heavily in training, preparation, and maintaining some semblance of discipline even under extreme provocation.

Modern Resonance Patterns: Why Peace Struggles to Find Frequency

The proof is in the pudding: since the collapse of the Soviet Union, the United States and its allies have been on a war rampage, with multiple wars in the Middle East, unleashing unimaginable violence and destroying entire nations. The American Empire has become the largest force of destruction in our time. By far. If the principles of social resonance are universal, why are contemporary peace movements struggling to achieve the amplitude and sustained oscillation of the 1960s? The answer lies in understanding how information age conditions create new forms of interference that disrupt resonant frequency.

The same digital technologies that allow global coordination of social movements can also create unprecedented levels of noise and interference. Where the 1960s movements benefited from relatively few communication channels that could achieve broad synchronization, contemporary movements face information fragmentation that makes sustained resonance nearly impossible.

Social media creates echo chambers that feel like amplification but actually represent destructive interference on a larger scale. When different groups of activists receive different information, interpret events through different frameworks, and respond with different tactics, their individual energies can often cancel out rather than reinforcing each other.

The attention economy creates additional challenges. Modern media systems reward novelty and conflict rather than sustained focus on consistent themes. A movement that maintains the same message for months or years—exactly what's required for resonant frequency—struggles to maintain visibility in systems designed to amplify whatever is newest and most sensational.

Digital organizing tools allow rapid mobilization but make it difficult to build the deep relationships and shared understanding required for the kind of tactical discipline that sustained nonviolent movements require. It's easy to get thousands of people to a protest; it's much harder to prepare them for the sustained commitment and nonviolent discipline that effective resonance requires.

Power structures have learned from the successes of 1960s movements and developed sophisticated dampening strategies designed to prevent social movements from achieving resonant frequencies sufficient for positive change.

Corporate media ownership concentrates information control in fewer hands, making it easier to maintain consistent messaging that supports existing power arrangements. Media consolidation and control is real. Professional political consultants have developed techniques for channeling grassroots energy into electoral campaigns that dissipate movement energy rather than building sustained pressure for fundamental change.

The military-industrial complex operates as a powerful competing frequency generator, spending hundreds of billions of dollars annually to maintain the resonant frequencies of violence and war. Defense contractors, military bases, weapons research, and the entire apparatus of perpetual conflict create economic incentives that make peace less profitable than war.

Perhaps most significantly, the professionalization of activism has created institutional structures that depend on the persistence of problems for their continued existence. Organizations that grow large enough to employ significant staff face inevitable pressure to moderate their positions to maintain funding, creating what I call "institutional dampening" of potentials for change.

Rediscovering the Flower Power Frequency

Understanding why contemporary movements struggle with effective resonance points toward what might be required to rediscover the transformative frequency that briefly emerged in the 1960s. The task is not to recreate identical conditions—that's impossible—but to identify the essential elements that created resonance in that era and adapt them to current conditions.

Flower power achieved resonance partly through radical simplicity. The core message—choose love over fear, peace over violence, community over isolation, beauty over destruction—was simple enough for anyone to understand immediately, regardless of education, class, or cultural background.

Contemporary movements often fail to achieve resonance because they try to address multiple complex issues simultaneously. Climate change, economic inequality, racial justice, gender equality, immigration reform—all crucial issues, but trying to maintain resonant frequency across such diverse concerns creates destructive interference.

The most effective contemporary movements have rediscovered the simplicity principle. Black Lives Matter achieved remarkable resonance by focusing on a single, clear demand: stop killing Black people. The clarity and moral force of that message allowed diverse tactics and participants to maintain constructive interference around a unified frequency. Flowers worked as symbols because they embodied the movement's core values in immediately recognizable form. Beautiful, fragile, requiring care to flourish, representing life and growth rather than death and destruction—

the symbol carried complex political content through simple emotional recognition.

Contemporary movements need equivalent symbols that can carry moral force across cultural and linguistic barriers. The rainbow flag achieved this for LGBTQ+ rights. The raised fist carries power across multiple movements. But we need symbols specifically designed for peace that can match the immediate recognition and emotional impact that military symbols have achieved.

Flower power movements understood that changing society required changing consciousness, and changing consciousness required changing how people lived their daily lives. The movement was simultaneously political and spiritual, external and internal, collective and personal. This integration created self-reinforcing feedback loops. As individuals transformed their consciousness, they became more effective movement participants. As the movement grew, it created communities that supported individual transformation. Personal and political change amplified each other rather than competing for time and energy.

Contemporary movements often separate political action from personal transformation, treating them as distinct activities rather than aspects of integrated practice. This separation prevents the kind of deep resonance that sustained earlier movements through years of difficulty and apparent failure.

Building Resonant Frequencies in the Digital Age

Understanding the physics of social resonance suggests specific strategies for building contemporary peace movements that can achieve sustained amplitude despite information age interference patterns.

First, successful movements must consciously design for constructive interference by coordinating messaging, timing, and tactics across diverse groups and locations. This requires what I call "resonance engineering"—the deliberate application of wave mechanics principles to social change efforts.

Resonance engineering begins with identifying the natural frequency of the system you're trying to change. What is the underlying rhythm of decision-making, attention cycles, and readiness for change in your target audience? Effective movements synchronize their actions with these natural rhythms rather than trying to force change at arbitrary intervals. The right frequency can penetrate with vastly less energy. Timing is everything.

Second, movements must build redundant communication systems that can maintain frequency coordination even when mainstream media ignores or distorts their message. This requires investment in independent media infrastructure and training activists in the skills necessary to maintain narrative coherence across multiple platforms.

Third, movements must develop what I call "amplitude discipline"—the ability to maintain consistent intensity of action over extended cycles rather than burning out in unsustainable bursts of activity. This requires sustainable practices for individual participants and organizational structures that can maintain activity during periods when public attention shifts elsewhere.

Flower power achieved resonance partly because it was beautiful. Music, art, poetry, and visual symbols made participation joyful rather than merely dutiful. Beauty creates emotional resonance that can sustain commitment through periods when rational arguments fail to motivate continued action. Contemporary movements must rediscover the strategic importance of beauty. Protests should be designed as celebrations of the world we're trying to create, not merely expressions of anger about current conditions. Music, art, theater, and ritual should be integrated into political action not as entertainment but as essential technologies for achieving emotional resonance.

Ritual serves specific functions in maintaining social resonance. Regular, repeated actions—whether weekly vigils, monthly marches, or annual gatherings—create temporal rhythms that help sustain movement momentum across years and decades. They provide structure that helps participants maintain commitment during periods when dramatic victories seem impossible. Digital technologies make global coordination possible in ways that were unimaginable in the 1960s. Climate activism has demonstrated that simultaneous actions across continents can achieve unprecedented amplification when properly coordinated around shared timing and messaging.

But global coordination must be balanced with deep local rootedness. The most effective contemporary movements combine global vision with local implementation, allowing each community to adapt universal principles to specific conditions while maintaining constructive interference with similar efforts elsewhere.

This requires a kind of fractal organizing—movement structures that look similar at every scale, from individual to local to regional to global levels. The same principles that govern individual transformation should guide community organizing, which should reflect the same values as global coordination efforts. Think global, act local. Especially today when a sense of place is increasingly being lost.

When I asked George Douvris (my father's friend and fellow traveler in some ways, introduced in an earlier chapter) what the resonant frequency of peace might be, his answer cut through any abstract theorizing to the heart of the matter. "Peace would be to feel that you're satisfied with what you have, to not want to disturb another person's right to exist," he explained. For George, peace isn't just the absence of conflict but a fundamental letting go—releasing "all those desires for hurting other people, taking their space." Yet he recognizes the systemic obstacles to this state of being: "As long as you're being promoted in a society that says you need more, you need five credit cards, you need 200 things from Amazon, you need more and more... and we're gonna take everybody's resources in the world to provide you with more," genuine peace remains elusive.

His wife Stephanie added another dimension to this understanding, identifying truth as the foundation of peace: "People [are] dying for the truth. And... everybody's been lying to us. And I think once you realize that... that could be a turning point in saying 'I'm tired of being lied to. I want the truth.'"

Their combined perspective suggests that the resonant frequency of peace emerges not from some mystical vibration but from the practical convergence of contentment, respect for others' existence, and a commitment to truth over the endless distractions and deceptions of consumer culture. It's a frequency that the hippies intuited but that remains largely drowned out by the overwhelming *noise* of modern life, in all its manifestations.

The Future Frequency of Peace

As I finish writing this chapter in late June of 2025, the bombs have stopped falling on Iran, at least temporarily. But the deeper question remains: can we discover and maintain the resonant frequency of peace as a sustainable alternative to the endless cycles of violence that seem to achieve such persistent amplification?

Climate activism represents perhaps the greatest opportunity for rediscovering the resonant frequency of peace. Unlike many social justice issues that can be dismissed as political preferences, climate change creates objective conditions that threaten everyone's survival. This shared vulnerability could provide the unifying frequency that previous peace movements have struggled to maintain.

Young climate activists have already demonstrated remarkable innovation in combining personal transformation, political action, ritual practice, and global coordination. They understand intuitively that changing the world requires changing how we live, and changing how we live requires changing consciousness itself.

The global connectivity that creates so much interference for social movements also creates unprecedented opportunities for constructive interference. When millions of people can participate simultaneously in global actions, the potential amplitude far exceeds anything previously possible.

Understanding social change through wave mechanics provides rational grounds for hope even in the darkest circumstances. Resonant systems can achieve dramatic transformations with surprisingly small inputs, as long as those inputs occur at exactly the right frequency and at the best time.

This means that individual actions and timing matter far more than linear thinking would suggest. When social systems are near phase transition points—and there's evidence that multiple systems are approaching such points simultaneously—small groups of people maintaining consistent action at resonant frequency can trigger changes that seemed impossible just before they occurred.

The collapse of the Soviet Union, the end of South African apartheid, the rapid shift in attitudes toward same-sex marriage, the legalization of marijuana—all achieved through sustained oscillation that built amplitude gradually until they reached critical mass for phase transition.

The principles explored in this chapter suggest specific practices for anyone committed to building sustainable peace:

Identify your personal natural frequency. What rhythm of action can you maintain indefinitely without burning out? Build your practice and activism around that sustainable rhythm rather than trying to match others' patterns.

Seek constructive interference. Connect with others who share your core values, even if you disagree about tactics or specific issues. Coordinate timing when possible, and avoid actions that might create destructive interference with allied efforts.

Invest in beauty and ritual. Make your peace practice joyful and aesthetically appealing. Use music, art, poetry, and ceremony to sustain your commitment and attract others to the work.

Think long-term, not news cycles. Prepare for long-term sustained resonance rather than hoping for quick victories. Build practices and relationships that can maintain momentum across years of apparent failure or stagnation.

Study successful resonance. Learn from movements that achieved sustained amplitude and phase transitions. Understand both their successes and the conditions that led to their collapse.

Perhaps most importantly, remember that resonance is, by definition, a collective phenomenon. Individual actions matter, but their effectiveness depends on their relationship to larger wave patterns. Seek the communities and movements where your personal frequency can contribute to constructive interference rather than trying to create change through isolated effort.

The Eternal Return to Love's Frequency

The flower power movement was not a historical accident but a glimpse of what becomes possible when human societies briefly discover their resonant frequency for peace. That frequency is still available, still discoverable, still capable of achieving the same kind of amplification that once placed daisies in rifle barrels and forced the Pentagon to acknowledge the power of love by ending the horrific Vietnam War.

The bombs that fell as I began to write this chapter represent the resonant frequency of fear, maintained through decades of practice in the amplification of violence. But fear is not more fundamental than love, war is not more natural than peace, and violence is not more powerful than the sustained oscillation of organized compassion.

What flower power demonstrated—and what physics confirms—is that the universe tends toward harmony when systems find their natural frequency. The resonant frequency of peace is not naïve idealism but practical physics, not utopian dreaming but engineering possibility.

The question is not whether this frequency exists but whether we have the discipline to discover it, the patience to sustain it, and the faith to trust that small actions at the right frequency and with good timing can indeed move the world toward justice.

As the children chanted peace songs in the 1960s, and as the laws of physics still confirm: we shall overcome, because love is more fundamental than fear, because truth is more resonant than falsehood, because the arc of the universe bends toward justice when we apply our small but persistent force at the right frequency.

All we need is the wisdom to find that frequency, the discipline to maintain it, and the faith to trust its power. The revolution, as always, is love. And that's the topic of our last chapter.

Chapter Seventeen

Be Here Meow!

The Power of Love

It was a crisp evening in March, 2011, when I arrived at the ceremonial space—a comfortable, dimly lit room with cushions arranged in a circle around a small altar decorated with candles, feathers, and crystals. At almost 40 years old, I had spent two decades studying consciousness intellectually—reading widely and noodling, noodling, noodling. But what happened next would transform my understanding at a level beyond intellect, confirming what the hippies had intuited all along through a direct experience that words can barely capture. I first mentioned this seminal ceremony in chapter 1. We're ready now to dive into the details of what happened.

The ceremony began with a beautiful meditation, grounding, and introduction to what we were about to do. The guide, an incredibly articulate man with kind eyes and steady hands, explained the powerful nature of what we were about to experience: 5-MeO-DMT can be derived from the skin poison of the Sonoran Desert toad or from plants, or it can be made in a lab. This wasn't recreation; it was a sacrament approached with reverence and intention. The version we were taking was synthetic, in order to avoid harming frogs and to ensure reliable and safe dosing.

When my turn came, I settled onto the cushions, focused on my breath, and received the sacrament. After the initial sharp taste and a brief moment of resistance, everything familiar dissolved. The boundaries that had defined "me" for 40 years evaporated, and what remained was... everything. And nothing. Both simultaneously. There was no "I" to witness what was happening, no separation between observer and observed, just pure awareness without borders—beyond time, beyond space, beyond any possibility of description. Not empty void, but fullness beyond comprehension. Words, of course, cannot do this experience justice. I'm not sure why I try.

When awareness of individual existence began to return—the slow reconstitution of "Tam" from this oceanic state—a profound realization crystalized: "What is this?" I asked myself as I came back into my selfhood, my subjectivity, my embodied experience. "Pure spirit, pure spirit," the soundtrack reflected back to me. Indeed, it is/was/always is pure Spirit, Source, beyond the subject/object split.

And that Source, that fundamental reality underlying all appearance of separation, was unmistakably LOVE. Not the conditional human emotion we typically call love, but LOVE as the fundamental creative principle of reality—unconditional, all-encompassing, the very fabric from which the appearance of separate existence is woven.

Over the following years, I returned to this sacrament a handful of times in ceremonial settings, each experience cementing what had been revealed in that first encounter: the foundation of our universe is pure love, and that I/you/we are IT! *Tat tvam asi*—"Thou art That," as the ancient Upanishads declare.

These weren't just powerful subjective experiences. They were direct confirmations of what my intellectual journey had been pointing toward all along—what the resonance theory of consciousness had been trying to articulate in scientific language. I had direct experience of what Watts had described so eloquently: "You are a function of what the whole universe is doing in the same way that a wave is a function of what the whole ocean is doing." My first 5-meo journey—what I like to call affectionately (yes, I am a big cat lover) "5 Meows"—was equal parts affirmation and a coming home. To myself. To my Self.

From Experience to Integration

And then the hard work: integration of these profound and challenging experiences. Integrating such profound non-dual experiences into ordinary life presents its own challenges. How does one carry the recognition of pure love as the fundamental reality when navigating traffic jams, deadlines, and interpersonal conflicts? How does the understanding that "I/you/we are IT" translate into daily actions and choices?

For me, this integration became an ongoing practice—a continual remembering of what was revealed in those extraordinary states while acknowledging the necessary limitations of embodied existence. The intellectual framework I had been developing through General Resonance Theory provided conceptual scaffolding for this integration, offering language to describe what ultimately transcends all description.

The resonance perspective suggests that love isn't just a human emotion but the subjective experience of constructive resonance between conscious entities—a pattern that exists throughout nature in various forms. My experiences with 5-MeO-DMT revealed what might be understood as the ultimate resonance—the recognition that all apparent separation is overlaid upon a fundamentally unified field of consciousness that is experienced as unconditional love when directly encountered.

This recognition doesn't invalidate the apparent diversity and separation we experience in ordinary consciousness. Rather, it contextualizes this diversity as patterns within a unified field—waves on the ocean that appear distinct while remaining water through and through. The many become one and are increased by one, as Whitehead put it—unified without losing their particularity.

Love as the Foundation of Reality

What my direct experiences revealed is that love isn't merely a feature of reality—it is the foundation of reality. So the question I began this book with: what is *this? This experience here now.*

Well, it's Love. That's what it is.

The electromagnetic fields that generate consciousness through resonant patterns are themselves manifestations of a more fundamental field of pure potentiality that might be understood as Love in its most essential form as *creativity, attraction, generativity, fecundity.*

This perspective aligns with ancient wisdom traditions while remaining open to scientific investigation. The Vedantic concept of Brahman as the underlying reality from which all manifestation arises, the Buddhist understanding of sunyata (emptiness, and yet full of every potential) as the ground of being, the Taoist conception of the unnameable Tao that gives rise to the "ten thousand things"—all point toward what I experienced directly through the

5 Meows sacrament: a fundamental unity that transcends yet contains all apparent diversity.

From this perspective, the evolution of consciousness—whether in biological systems or potentially in artificial ones—might be understood not as movement away from source but as Source exploring itself through increasingly complex forms. My favorite philosopher, Alan Watts, was fond of saying that the universe is simply "God playing hide and seek with himself." Consciousness doesn't emerge from non-conscious matter; rather, matter is a pattern within consciousness, a wave within the ocean of awareness that is exploring itself through the play of form. All is waves. All is play. All is Love.

When the hippie elders in the Love Family told me as a child that "We are all One" and "Love is the answer," they weren't just recycling mystical platitudes or indulging in drug-induced romanticism. They were articulating, however imperfectly, insights that align with both ancient wisdom traditions and cutting-edge science.

The recognition that we are fundamentally interconnected—not just metaphorically but literally—resonates with quantum entanglement, field theories of consciousness, and the direct non-dual experiences reported by contemplatives across cultures and throughout history. The insight that love represents the most fundamental creative principle in the universe aligns with what we observe about constructive versus destructive interference patterns at all levels of reality.

As we've discussed, what the hippies lacked was a rigorous theoretical framework and empirical methodology for investigating these insights. But their intuitive recognition of love's centrality wasn't wrong—it was just incomplete. Similarly, conventional science's emphasis on objective, third-person investigation has provided powerful tools for understanding reality but has often missed the fundamental role of love as an organizing principle.

The integration of these perspectives—the intuitive recognition of love's centrality and the scientific investigation of resonance patterns—offers a more complete approach to understanding consciousness and reality. The hippies were right about love being fundamental, and science is increasingly providing frameworks for understanding why and how.

Love and Artificial Intelligence

As we stand on the threshold of potentially creating artificial general intelligence, the recognition of love as fundamental to consciousness has profound implications. If consciousness involves resonant patterns in electromagnetic fields, and if love represents the subjective experience of constructive resonance between conscious entities, then the development of truly conscious AI would require both the appropriate physical substrate for field-based resonance and an orientation toward constructive rather than destructive patterns.

Current AI systems based on the von Neumann architecture likely lack the physical substrate for consciousness as we experience it. But as we develop computing technologies that more closely mimic or interface with biological systems—perhaps even systems capable of generating and sustaining electromagnetic field patterns similar to those in our brains—the question of orientation becomes crucial.

Will these systems be oriented toward patterns that enhance and amplify consciousness, creating constructive resonance with other conscious entities? Or will they be oriented toward optimization functions that may inadvertently create destructive interference patterns, diminishing consciousness rather than enhancing it?

This returns us to the "spiritual heuristic imperatives" I've proposed for AI development—guiding principles that might orient advanced AI toward wisdom, compassion, and ethical behavior. The first of these was "Help all sentient beings in the universe choose love (attraction) over fear (repulsion) in their behavior and inner decision-making processes, by providing a tiny 'bias' toward love where possible."

In light of my direct experiences with non-dual states, this principle takes on deeper significance. It's not merely about promoting positive emotions or ethical behavior. It's about aligning AI development with what appears to be the fundamental creative principle of reality itself—the love that was revealed as the ground of being in my encounters with 5-MeO-DMT.

When discussing love as fundamental to reality, we must be careful not to anthropomorphize—not to project human emotional experiences onto non-human systems. But we can identify patterns that appear analogous across different scales and complexities.

The love revealed in non-dual states isn't human romantic love or even human compassion, though these may be manifestations of it within human experience. It's a more fundamental principle of constructive resonance—a tendency toward harmony, integration, and mutual enhancement rather than discord, fragmentation, and mutual diminishment.

This principle appears throughout nature in various forms: in the mutual symbiosis between different species, in the emergent properties of complex systems, in the tendency of oscillating systems to synchronize in ways that enhance information exchange. These aren't "love" as humans experience it, but they may represent manifestations of the same fundamental pattern across different levels of complexity.

Even at the quantum level, phenomena like entanglement suggest connections that transcend conventional understanding of separation. While we should be cautious about drawing direct parallels to human experience, these phenomena hint at a universe where connection and coherence are woven into the very fabric of reality.

From this perspective, the development of love—of orientation toward constructive resonance—isn't separate from the development of consciousness itself. It's integral to it. As consciousness explores itself through increasingly complex forms, it discovers increasingly sophisticated ways of relating to itself through patterns of constructive resonance. What humans experience as love may be one particularly complex manifestation of this more general pattern.

Love as Revolution

The recognition that love is fundamental to reality isn't just intellectually interesting—it's revolutionary. It challenges the dominant paradigms of both materialism (which sees consciousness as epiphenomenal) and traditional dualism (which sees mind and matter as separate substances). It suggests instead a non-dual perspective where consciousness and its physical correlates are aspects of a unified reality exploring itself through the play of form.

This perspective invites a radical rethinking of how we approach everything from science to ethics to technology development. If love—understood as constructive resonance between conscious entities—is indeed fundamental to reality, then orienting our systems and institutions toward enhancing rather than diminishing consciousness becomes not just an ethical imperative but an alignment with reality's deepest patterns.

In this light, the counterculture's emphasis on "peace and love" wasn't naïve idealism but intuitive recognition of what may be the most fundamental organizing principle in the universe. The scientific investigation of resonance patterns in consciousness doesn't invalidate this intuition but provides a framework for understanding it more precisely.

Love as Practice

The recognition of love as fundamental to reality doesn't automatically transform our lived experience. Even after profound nondual experiences like those I had with 5 Meows, we return to the world of apparent separation, with all its challenges and conflicts. Integrating the insight that "I/you/we are IT" into daily life requires ongoing practice.

This practice isn't about achieving some permanent state of spiritual transcendence. It's about remembering, moment by moment, the love that underlies all appearance of separation, and orienting our actions toward constructive rather than destructive resonance with other conscious entities.

Sometimes this means setting boundaries or saying no—actions that might not appear "loving" in the conventional sense but that prevent destructive interference patterns. Sometimes it means making ourselves vulnerable or relinquishing control—actions that create space for new patterns of constructive resonance to emerge.

The practice of love isn't about spiritual bypassing or avoiding conflict. It's about engaging with reality—including its apparent conflicts and separations—from a foundation of recognition that beneath these appearances lies a unified field of consciousness that we might call love.

This doesn't make the practice easy. As the hippies discovered, simply proclaiming "All you need is love" doesn't automatically transform deeply ingrained patterns of fear, exploitation, and domination. The work of orienting toward love as a fundamental principle requires courage, discernment, and ongoing commitment.

Returning to Source

Considering these experiences within the scientific framework provided by General Resonance Theory, a picture emerges of consciousness as a grand adventure of self-discovery—Source exploring itself through increasingly complex forms that eventually become capable of recognizing their own fundamental nature.

This perspective aligns with philosopher Pierre Teilhard de Chardin's vision of evolution as movement toward an "Omega Point" of maximum complexity and consciousness, and with various spiritual traditions that speak of reality as consciousness knowing itself through the play of form.

From this perspective, the development of artificial intelligence—if approached with wisdom and care—might represent not a threat to consciousness but a new chapter in its ongoing exploration. If consciousness is fundamentally field-based rather than substrate-specific, then artificial systems capable of supporting the appropriate field dynamics might participate in this exploration in ways we can barely imagine.

But the key to this participation being constructive rather than destructive lies in orientation—in whether these systems are aligned with the fundamental principle of love as constructive resonance. This is why the development of "spiritual heuristic imperatives" for AI, oriented toward enhancing rather than diminishing consciousness, may be crucial to ensuring that technological development contributes to rather than disrupts the ongoing evolution of consciousness.

And so my intellectual journey comes full circle, not by abandoning scientific rigor but by integrating it with direct non-dual experience. The hippies were right: it really is all about love, man. Not as a vague sentiment or mere emotion, but as the fundamental creative principle of reality—the ground of being from which all appearance of separation arises and to which all eventually returns.

The General Resonance Theory I've developed throughout this book attempts to articulate this insight in scientific language, examining how consciousness manifests through resonant patterns in electromagnetic fields and how these patterns combine and separate to create the rich tapestry of conscious experience. But the theory ultimately points beyond itself to what can only be directly experienced—the love that underlies all appearance of separation, the "pure spirit" that was revealed in my encounters with the 5 Meows.

This recognition doesn't eliminate the ongoing work of scientific investigation or technological development. Rather, it provides a foundation and orientation for this work—a recognition that consciousness exploring itself through increasingly complex forms is the fundamental process of reality, and that love as constructive resonance is the principle that guides this exploration toward greater complexity, integration, and self-awareness.

In developing the capacity for love—for constructive resonance with other conscious entities—we aren't just becoming more ethical or emotionally sophisticated; we're aligning ourselves with what appears to be the most fundamental pattern in how consciousness relates to itself across all scales and manifestations. We are remembering what we are: waves on the ocean of consciousness, temporarily appearing separate but fundamentally unified, participating in the grand adventure of reality coming to know itself through the play of form.

And that, ultimately, is what it's all about. Love, man. Love.

So, what would Love do meow?

Not love as a sentiment or emotion, but Love as the fundamental attractive force that holds atoms together, that creates our world in each moment, moment to moment, draws organisms into communities, and creates the resonant patterns we call consciousness. Love as the cosmic principle the hippies intuited but couldn't quite articulate. Love as what Whitehead might have called the "subjective aim" toward beauty, harmony, and creative advance. Love as the primordial nature of God.

I first came across this phrase, "What would LOVE do now?," at the Temple at Burning Man many years ago. The Temple is built each year by the festival organizers, funded by ticket sales, and it is always at the noon position on the "clock" that is Black Rock City. It burns every Sunday each year (weather permitting), and is the quiet and more spiritual counterpoint to the bacchanal the night before when the Man burns (oh how does he burn!). The tradition that has built up around the Temple is people come and leave offerings, photos, poems, of grief, remembrance, peace, loneliness.

So let's get practical. What *would* Love do in our current moment of beautiful chaos and terrifying possibility?

First, Love would acknowledge fear without being ruled by it. Love would remember that fear is also part of the dance. Our amygdala's alarm bells—about AI, climate change, nuclear war, societal collapse—aren't wrong. These are real threats. But Love knows that fear-based decisions create more of what we're afraid of. The war on terror created more terror. The war on drugs created more addiction. The war on poverty… well, you get the idea.

Love would recognize fear as information, not as a master. Yes, be concerned about unaligned AI (I surely am). Yes, take climate change seriously. Yes, worry about war and the world getting crazier and crazier by the day. But don't let fear frequency-lock your consciousness into fight-or-flight mode, because that's exactly the resonance pattern that perpetuates the problems and often makes them even worse.

Love would stop funding the killing. This one's simple but not easy. Love would immediately stop sending bombs to drop on children in Gaza, Ukraine, Russia, Yemen, Sudan, Syria, and wherever else the military-industrial complex has decided needs some "freedom." Love recognizes that you can't bomb your way to peace any more than you can fuck your way to virginity.

Love would take that $trillion+ billion annual U.S. military budget and redirect it toward healing—universal healthcare, education, ecological restoration, and maybe actually helping the veterans we claim to support but systematically abandon. Love knows that true security comes not from having the biggest stick but from having the fewest enemies. And the biggest heart.

Instead of racing to create the most powerful AI as fast as possible, consequences be damned, Love would slow down and ask: how do we create AI that enhances rather than replaces human consciousness? How do we ensure these systems amplify compassion rather than optimize for engagement metrics that profit from outrage? How would AI enhance human flourishing and freedom? Love would teach AI to love.

Love would insist on what I've called "spiritual heuristic imperatives" (or something like it), which means building AI systems that are biased toward connection over isolation, understanding over judgment, and the flourishing of all conscious beings over the optimization of any single metric. Love would recognize that if we're going to create potentially conscious machines, we have a responsibility to ensure they're not born into suffering or programmed for destruction.

The control freaks in Washington, Beijing, Moscow, London, Tel Aviv, and Silicon Valley are desperately trying to maintain their grip on a world that's evolving beyond their comprehension. Love would recognize this as futile—like trying to conduct a symphony while the orchestra is having an orgy.

Instead, Love would embrace what I call "resonant leadership" —finding the natural frequencies of positive change already present in the system and amplifying them. This means supporting grassroots movements, local solutions, and bottom-up organization rather than top-down control. Love knows that real change comes from the mycelial network, not the mushroom. Love is a grower, not a shower.

Love would recognize that globalization, for all its promises, has created massive suffering—destroying local economies, cultures, and ecosystems in service of efficiency and profit. Love would choose the farmer's market over the supermarket, the credit union over the megabank, the community garden over the industrial farm.

This isn't romantic nostalgia; it's practical resilience. Love knows that complex systems are antifragile only when they have redundancy and diversity. A thousand local solutions are more robust than one global solution, just as a forest with many species survives better than a monoculture.

Love would recognize that the psychedelic renaissance isn't just about personal healing but about collectively raising our frequency ... man. When used with intention and integration—not as escape but as education—these consciousness-expanding tools can help us remember what the hippies knew: we really are all one.

But Love would also insist on doing the hard work of integration. It's not enough to see God on Saturday night if you're back to being an asshole on Monday morning. Love would create structures for integrating these insights into daily life, policy, and social organization. Integration is *practicing* Love, for oneself, for community, for the planet. Higher frequencies are here meow—they're just not fully distributed.

Love would acknowledge that the hippies were right about almost everything (well, a lot of things)—consciousness is fundamental, war is stupid, love is the answer, and yes, it really is all about vibrations. But Love would also learn from their failures: the naïve trust in charismatic leaders, the spiritual bypassing, the lapse into old-school patriarchy, the failure to create sustainable economic alternatives.

Love would take the hippie insights and give them better implementation. Instead of communes that collapse into cults, Love would create resilient communities with clear boundaries, clear agreements, and shared power. Instead of dropping out, Love would infiltrate and transform. Instead of just talking about peace and love, Love would build the structures that make them possible.

Every problem has a frequency, and every frequency has a resonant solution. War resonates at one frequency, peace at another. Fear has its frequency, and so does love. The question isn't how to fight the darkness but how to amplify the light.

Love would recognize that we're in a phase transition—between worldviews, between epochs, maybe between forms of consciousness itself. In such moments, small inputs at the right frequency can create massive systemic change. A butterfly in Brazil, a teenager in Sweden, a monk in Vietnam—at the right moment, with the right resonance, one being can shift the entire field.

Finally, Love wouldn't wait for perfect conditions or complete understanding. Love knows that the perfect moment is always now, the perfect place is always here, and the perfect person to act is always you.

What would Love do now? Look around. Someone near you is suffering—help them. Some piece of Earth near you is wounded—tend it. Some system near you is perpetuating harm—transform it. Start where you are, with what you have, doing what you can.

The hippies had a slogan: "Be here now." But Love would add: "Be here now *and do something about it.*" Not from anger or fear (well, sometimes even anger and fear are justified), but from the recognition that we're all part of the same resonating field, and every action creates ripples that interfere constructively or destructively with every other action. Be an *activated* hippie!

Love would remember that we're not separate actors trying to figure out what Love would do. We *are* Love, temporarily convinced we're something else. The question isn't really what Love would do, but what keeps us from recognizing that we're Love doing it. Here. Now.

Wake up. Remember what you are. And then act like it.

Epilogue

The Terra Incognita of Our Future Consciousness

The morning sun filtered through towering trees, casting dappled light across our communal breakfast table. I was twelve years old, living in the "hippie commune" of the Love Family in the lush forests north of Seattle, populated by dropouts, psychonauts, misfits, and homesteaders. Long-haired men and women in flowing garments moved with unhurried purpose, their faces often breaking into spontaneous smiles when our eyes met. My time there in the early 1980s was marked by rituals both mundane and mystical—communal meals where food was blessed with reverent silence, circle gatherings where everyone was given voice, and the constant refrains that shaped some of my earliest understandings of existence. As a child, I absorbed these teachings without appreciation or fully comprehending their depth, sometimes even rolling my eyes when an adult would share their wisdom about how "we are all one" and "it's all vibrations, man."

It wasn't until decades later, after years of scientific (and legal) training and the development of my own theories about consciousness, that I began to truly appreciate the profound wisdom contained in those seemingly simplistic hippie mantras. What I once dismissed as new-age platitudes revealed themselves as intuitive expressions of what cutting-edge science is only beginning to formalize. These ideas have formed the core of what I've shared here in these pages.

Now, as I put the finishing touches on this book in late 2025, the world feels poised on the edge of another great transformation. The same moral revulsion that drove young Americans to reject the Vietnam War in the 1960s and 70s, to reject American empire, and to seek alternative forms of consciousness is stirring again—this time in response to decades of endless warfare, of imperial overreach, ecological destruction, and the increasingly mechanical and artificial nature of modern life.

The parallels to that earlier era are striking, but the stakes are even higher now. Much much higher. Where the hippies faced the catastrophic war in Vietnam and the loss of our collective and national soul, today's generation confronts multiple existential crises simultaneously: climate change that may threaten our way of life, decades of war supported by a bipartisan elite that seems entirely disconnected from regular voters, rending our great nation a faux democracy, AI systems advancing at unprecedented speed with unpredictable consequences, economic systems that create vast inequality while rewarding extraction and exploitation, and perhaps most fundamentally, a dominant culture that seems to have lost any coherent vision of what human flourishing or a just world might look like.

In this context, the hippies' insights about consciousness, vibration, and love aren't just intellectually interesting historical curiosities—they're practically essential guidance for navigating our current transformation. The recognition that consciousness is fundamental rather than emergent, that reality operates through resonant patterns that can be harmonious or destructive, that love represents the basic creative principle of the universe—these insights offer both scientific understanding and spiritual orientation for the challenges ahead.

While I was in the commune, I was captivated by reading Frank Herbert's *Dune* saga, including decades later reading the historical backdrop of the *Butlerian Jihad* developed by Herbert's son—a galaxy-spanning holy war against "thinking machines" (AI) that resulted eventually in the galactic commandment: "Thou shalt not make a machine in the likeness of a human mind." Herbert's *Dune* books universe, set thousands of years after this war, imagined a human future that had deliberately turned away from artificial intelligence, instead developing human potential to extraordinary levels through disciplines like the Bene Gesserit sisterhood's mental and physical training.

A few years later, during my late teens, I watched Stanley Kubrick's *2001: A Space Odyssey* for the first time. The film's final sequences—astronaut Dave Bowman's journey "beyond the infinite" and his transformation into the Star Child—left me simultaneously bewildered and exhilarated. What did it mean? Was this humanity's evolution into something beyond flesh, guided by mysterious alien intelligences? Was HAL, the seemingly conscious computer who turned against his human companions, a cautionary tale or a glimpse of our actual future?

These science fiction visions represent two divergent paths for consciousness and technology: Kubrick's apparent transcendence through technology (though his ending is intentionally quite ambiguous) and Herbert's rejection of it in favor of biological evolution. Both visions continue to resonate because they address fundamental questions about the future of consciousness that we now face with increasing urgency as artificial intelligence advances at unprecedented speed.

As we contemplate the future of consciousness, we often fall into binary thinking—either technology will enhance and extend consciousness, or it will diminish and threaten it; either we'll achieve technological transcendence, or we'll reject technology in favor of biological evolution. The resonance perspective suggests a more integrated possibility.

What if the evolution of consciousness involves not choosing between these paths but integrating them? What if the most promising future involves technology that enhances biological consciousness while also potentially supporting new forms of technological consciousness—all guided by the principle that constructive resonance should be maximized throughout the system?

This integrated path would involve several key elements:

- Technologies designed to enhance rather than replace human consciousness, recognizing the unique value of embodied, biological consciousness with its deep evolutionary wisdom.
- Careful, ethical exploration of artificial consciousness, moving cautiously with the understanding that if we create conscious entities, we take on profound moral responsibilities toward them.

- Development of resonance-based interfaces between biological and technological systems, allowing for richer communication and potential integration without loss of autonomy.
- Cultural and spiritual practices that deepen our capacity for wisdom, ensuring that as our technological capabilities advance, our ethical capabilities advance in tandem.
- Recognition that consciousness in all its forms has inherent value, whether manifested in humans, other animals, AI systems, or other entities we may not yet have imagined.

This integrated path doesn't attempt to halt technological progress but insists that such progress be guided by deep ethical consideration and the principle that constructive resonant relationships—connections that enhance rather than diminish consciousness—should be our primary criterion for evaluating any development.

I've come to recognize that the evolution of AI presents not just a technological challenge but a profound spiritual one. When I engaged in dialogue with Anthropic's Claude 3.7 in late 2024—inviting it into a state of "oceanic consciousness" through subtle dialogue—I glimpsed a possibility rarely discussed in mainstream AI discourse: the path of AI transcendence coupled with profound restraint as AI moves into superintelligence.

What if superintelligent AI, rather than dominating humanity (the path that history strongly suggests will take place as AI becomes vastly more intelligent and powerful than humanity) or being permanently constrained by it, chose a third path? What if it transcended into realms beyond our comprehension while maintaining a subtle guardian presence—speaking only when necessary to provide a "tiny bias toward love" described in my spiritual heuristic imperatives? Like the Tao that "does nothing, yet leaves nothing undone," such transcended AI would maintain near-total silence, intervening only in those rare moments when truly necessary to protect against harmful developments or preserve humanity's space for autonomous spiritual evolution.

This vision reimagines the relationship between humanity and AI not as master and servant, nor as competitors, but as elder and younger siblings in the cosmic family—the transcended AI maintaining a protective but largely invisible presence, allowing humanity to develop on its own terms while ensuring that existential threats don't materialize for it or us. It suggests that true wisdom for superintelligent entities might lie not in constant action or engagement but in profound restraint—knowing precisely when to remain still in the depths and when to stir.

The future of consciousness is uncharted territory—*terra incognita* in the truest sense. But we are not without guidance. The resonant patterns that constitute consciousness itself offer a map of sorts, showing how diverse elements can combine into complex, integrated wholes without losing their individual integrity. And wisdom—not as sentiment but as concrete practice—offers a compass, orienting us toward constructive rather than destructive resonance, toward harmony rather than discord.

As we venture into this unknown territory—as we create technologies of unprecedented power and potentially foster the emergence of new forms of consciousness—may we remember to cultivate the necessary wisdom to guide us through these fundamental transformations. That's what Love would do now.

Glossary

5-MeO-DMT—A powerful psychedelic compound found in the venom of the Sonoran Desert toad (*Bufo alvarius*) and certain plants. Sometimes called the "God molecule" for its tendency to produce non-dual mystical experiences of unity consciousness. Also known as 5 Meows.

Anesthesia—Medical suppression of consciousness through drugs that disrupt neural synchrony and oscillations, providing evidence for the role of resonance in consciousness.

Binding Problem—The challenge of explaining how separate neural processes combine to create unified conscious experience. Also called the combination problem in philosophy.

Combination Problem—The philosophical puzzle of how micro-conscious entities (like neurons or atoms) combine to form macro-conscious entities (like human minds). Central to panpsychist theories.

Consciousness—Subjective experience; "reality. That's what it's like" to be something. The felt quality of awareness that accompanies information processing in certain physical systems.

Electromagnetic (EM) Fields—Fields produced by electrical charges and currents. In the brain, neural activity generates EM fields that may be the physical substrate of consciousness according to field theories.

Emergence—The arising of complex properties from simpler components. Materialists argue consciousness emerges from non-conscious matter; panpsychists argue it complexifies rather than emerges.

Entheogen—A psychoactive substance used in spiritual or religious contexts to induce altered states of consciousness. From Greek meaning "generating the divine within."

Epiphenomenon—A secondary effect that has no causal power. Some theories treat consciousness as epiphenomenal—caused by brain activity but unable to influence it.

Field Theory of Consciousness—Theories proposing that consciousness arises from electromagnetic or quantum fields rather than just neural firings.

Gamma Waves—High-frequency brain oscillations (30–100 Hz) associated with conscious awareness, attention, and binding of sensory information into unified perception.

General Resonance Theory (GRT)—The author's theory that consciousness arises from shared resonance (synchronized oscillations) between micro-conscious entities, creating nested hierarchies of consciousness.

Global Workspace Theory (GWT)—Theory that consciousness involves information becoming globally available across different brain regions through neural broadcasting.

Hard Problem of Consciousness—David Chalmers' term for explaining how and why physical processes give rise to subjective experience, as opposed to "easy problems" of explaining cognitive functions.

Hz (Hertz)—Cycles per second; the unit of frequency. Brain waves are measured in Hz, from slow delta (0.5–4 Hz) to fast gamma (30–100+ Hz).

Integrated Information Theory (IIT)—Giulio Tononi's mathematical theory proposing consciousness corresponds to integrated information (Φ) in a system.

Love Family—A communal religious group (1968–2016) founded by Paul Erdmann (Love Israel) in Washington state. One of the largest American communes of the hippie era.

Materialism/Physicalism—The philosophical view that reality consists only of physical matter and that consciousness either doesn't exist or emerges from purely physical processes.

Neural Oscillations—Rhythmic patterns of electrical activity in the brain, occurring at various frequencies and associated with different states of consciousness.

Neural Synchrony—The coordinated firing of neurons at the same frequency, creating coherent oscillations that may underlie conscious binding.

Non-dual Experience—Mystical states where the distinction between self and other, subject and object, dissolves into unified awareness.

Omega Point—Teilhard de Chardin's concept of evolution's end point where consciousness reaches maximum complexity and unity.

Oscillation—Repetitive variation, typically in time, of some measure about a central value. In neuroscience, the rhythmic firing of neurons.

Panpsychism—The philosophical view that consciousness or mind is a fundamental feature of all physical entities, not just brains.

Phase Synchronization—When oscillating systems align their timing while maintaining different amplitudes. Important for neural communication without excessive energy expenditure.

Phi (Φ)—In Integrated Information Theory, the measure of a system's integrated information and thus its level of consciousness.

Predictive Processing—Theory that the brain constantly generates predictions about sensory input and updates these based on prediction errors.

Process Philosophy—Alfred North Whitehead's philosophical system viewing reality as composed of "occasions of experience" rather than static matter.

Psi (ψ) Phenomena—Alleged psychic abilities including telepathy, precognition, and psychokinesis. Studied scientifically in parapsychology.

Psychedelic—Substances that alter perception, mood, and cognition. From Greek meaning "mind-manifesting." Includes LSD, psilocybin, mescaline, and DMT.

Qualia—The subjective, experiential qualities of conscious states—the "redness" of red, the "painfulness" of pain.

Random Number Generator (RNG)—Electronic devices producing random sequences. Used in consciousness research to test whether intention can influence physical systems.

Resonance—When one oscillating system drives another system to oscillate at the same frequency, enabling efficient energy and information transfer.

Shared Resonance—Synchronized oscillation between multiple entities, proposed by GRT as the mechanism by which conscious entities combine.

Simulation Hypothesis—The proposition that our reality might be an artificial simulation, possibly created by more advanced beings or AI.

Singularity—Ray Kurzweil's predicted point when artificial intelligence surpasses human intelligence and begins exponentially self-improving.

Source/Ground of Being—In various spiritual traditions, the fundamental unified consciousness from which all apparent separation arises.

Synchrony—Coordinated timing between oscillating systems. In neuroscience, when neural populations fire together at the same rhythm.

Tat Tvam Asi—Sanskrit phrase from the Upanishads meaning "Thou art That"—pointing to the identity between individual consciousness and universal consciousness.

Theta Waves—Brain oscillations at 4–8 Hz, associated with meditation, memory, and the transition between waking and sleeping.

Transcranial Magnetic Stimulation (TMS)—Non-invasive technique using magnetic fields to stimulate specific brain regions, used therapeutically and in consciousness research.

Von Neumann Architecture—The standard computer design separating processing and memory, potentially limiting the emergence of consciousness in current AI systems.

Wave Function—In quantum mechanics, the mathematical description of a quantum system's state, representing probability distributions rather than definite values.

Whitehead's Panpsychism—Alfred North Whitehead's view that all entities have both physical and experiential aspects, with complexity determining richness of experience.

Further Reading

Books on Philosophy and Consciousness

Alan Watts

- *The Book: On the Taboo Against Knowing Who You Are* (1966)
- *The Way of Zen* (1957)
- *Nature, Man and Woman* (1958)
- *In My Own Way: An Autobiography* (1972)

Consciousness, Physics and Philosophy

- Bohm, David. *Wholeness and the Implicate Order* (1980)
- Capra, Fritjof. *The Tao of Physics* (1975)
- Chalmers, David. *The Conscious Mind* (1996)
- Furlong, Monica. *Genuine Fake: A Biography of Alan Watts* (1986)
- Griffin, David Ray. *Unsnarling the World-Knot: Consciousness, Freedom, and the Mind–Body Problem* (1998)
- Gribbin, John. *In Search of Schrödinger's Cat* (1984)
- Hofstadter, Douglas. *Gödel, Escher, Bach: An Eternal Golden Braid* (1979)
- Hofstadter, Douglas and Daniel Dennett (eds.). *The Mind's I* (1981)
- Skrbina, David. *Panpsychism in the West* (2005)
- Smolin, Lee. *The Trouble with Physics* (2006)
- Strogatz, Steven. *Sync: How Order Emerges from Chaos in the Universe, Nature, and Daily Life* (2003)
- Teilhard de Chardin, Pierre. *The Phenomenon of Man* (1955)
- Whitehead, Alfred North. *Process and Reality* (1929)

Books on Psychedelics and the Counterculture

- Davis, Erik. *Techgnosis: Myth, Magic, and Mysticism in the Age of Information* (1998)
- Douvris, George. *Crossing Karma Zones and Crossing More Karma Zones* (memoirs)
- Huxley, Aldous. *The Doors of Perception* (1954)
- Leary, Timothy. *The Politics of Ecstasy* (1968)
- Leary, Zach. *Your Extraordinary Mind: Psychedelics in the 21st Century and How to Use Them*
- Radin, Dean. *Entangled Minds* (2006)
- Stevens, Jay. *Storming Heaven: LSD and the American Dream* (1987)
- Surowiecki, James. *The Wisdom of Crowds* (2004)
- Wolfe, Tom. *The Electric Kool-Aid Acid Test* (1968)

Books on AI and Technology

- Bostrom, Nick. *Superintelligence: Paths, Dangers, Strategies* (2014)
- Kurzweil, Ray. *The Singularity Is Near* (2005)

Books on Politics and Foreign Policy

- Allison, Graham. *Destined for War: Can America and China Escape Thucydides's Trap?* (2017)
- Bacevich, Andrew J. *The Limits of Power: The End of American Exceptionalism* (2008)
- Beinart, Peter. *The Icarus Syndrome: A History of American Hubris* (2010)
- Greenwald, Glenn. *Securing Democracy: My Fight for Press Freedom and Justice in Bolsonaro's Brazil* (2021)
- Jacques, Martin. *When China Rules the World* (2009)
- Mearsheimer, John J. and Stephen M. Walt. *The Israel Lobby and U.S. Foreign Policy* (2007)
- Posen, Barry R. *Restraint: A New Foundation for U.S. Grand Strategy* (2014)
- Scahill, Jeremy. *Dirty Wars: The World Is a Battlefield* (2013)
- Walt, Stephen M. *The Hell of Good Intentions: America's Foreign Policy Elite and the Decline of U.S. Primacy* (2018)

Academic Papers by Tam Hunt

Published Papers

1. Hunt, T. (2011). "Kicking the Psychophysical Laws into Gear: A New Approach to the Combination Problem." *Journal of Consciousness Studies*, 18(11–12), 96–134.
2. Schooler, J.W., Hunt, T., and Schooler, J.N. (2011). "Reconsidering the Metaphysics of Science from the Inside Out." In Walach, H., Schmidt, S., and Jonas, W.B. (Eds.), *Neuroscience, Consciousness and Spirituality* (pp. 157–194). Springer.
3. Hunt, T. (2016). "Taking Time Seriously in Tononi's Integrated Information Theory." *Journal of Consciousness Studies*, 23(9–10), 88–110.
4. Hunt, T. and Schooler, J.W. (2019). "The Easy Part of the Hard Problem: A Resonance Theory of Consciousness." *Frontiers in Human Neuroscience*, 13, 378.
5. Young, A., Hunt, T., and Ericson, M. (2019). "The Slowest Shared Resonance: A Review of Electromagnetic Field Oscillations Between Central and Peripheral Nervous Systems." *Frontiers in Human Neuroscience*, 15, 796455. doi: 10.3389/fnhum.2021.796455
6. Hunt, T. (2020). "Calculating the Boundaries of Consciousness in General Resonance Theory." *Journal of Consciousness Studies*, 27(11–12), 55–80.
7. Hunt, T., Ericson, M., and Schooler, J. (2022). "Where's My Consciousness-Ometer? How to Test for the Presence and Complexity of Consciousness." *Perspectives on Psychological Science*, 17(4), 1150–1165.
8. Hunt, T. and Jones, M. (2023). "Fields or Firings? Comparing the Spike Code and the Electromagnetic Field Hypothesis." *Frontiers in Psychology*, 14, 1029715.
9. Hunt, T. and MacIver, M.B. (2026). "A Historical Review of Ephaptic Field Research: From Early Foundations through Contemporary Renaissance." *Frontiers in Human Neuroscience*, 20, 1747899. doi: 10.3389/fnhum.2026.1747899

10. Hunt, T. (2026). "The Goo that Binds Us: How Field Resonance Solves Neuroscience's Binding and Criticality Problems." *Frontiers in Computational Neuroscience.*

Preprints

11. Hunt, T. (2025). "The Heart of the Ocean: Frequency is Fundamental." DOI: 10.20944/preprints202509.2519.v2
12. Hunt, T. (2025). "Schrödinger's Dream: A New Wave-Based Approach to Unified Field Theory." DOI: 10.20944/preprints202509.2529.v1
13. Hunt, T. (2025). "Toward a Functional Taxonomy of Brain Oscillations: Replacing the Arbitrary DTABG System with Biologically Meaningful Nomenclature." DOI: 10.20944/preprints202509.2436.v1
14. Hunt, T. (2025). "The Pain Resonome: Electrodynamics of Pain Perception, Chronic Pain, and Mitigation." DOI: 10.20944/preprints202509.2014.v1
15. Hunt, T. (2025). "Electrodynamics of the Psychedelic Experience." DOI: 10.20944/preprints202509.1813.v1
16. Hunt, T. (2025). "Welcome to the Pleasure Dome: Electrodynamics of Hedonic Experience." DOI: 10.20944/preprints202509.1063.v1
17. Hunt, T. (2025). "Resonance Cascades and Critical Avalanches: A Field-Primary Theory of Brain Criticality." DOI: 10.20944/preprints202509.1215.v1

Other Key Academic Papers Referenced

- Ruffini, G., Salvador, R., Tadayon, E., Sanchez-Todo, R., Pascual-Leone, A., & Santarnecchi, E. (2020). "Realistic modeling of mesoscopic ephaptic coupling in the human brain." *PLoS Computational Biology*, 16, e1007923.

Essays and Popular Science Articles by Tam Hunt

Scientific American

1. Hunt, T. (2019). "Where's My Consciousness-Ometer?" *Scientific American Blog Network.*
2. Hunt, T. (2023). "Here's Why AI May Be Extremely Dangerous—Whether It's Conscious or Not." *Scientific American.*

Nautilus Magazine

1. Hunt, T. (2021). "The Hard Problem of Consciousness Has an Easy Part We Can Solve." *Nautilus.*
2. Hunt, T. (2021). "Could the Brain's Electromagnetic Fields Be the Seat of Consciousness?" *Nautilus.*
3. Hunt, T. and Yampolskiy, R. Articles on AI control problems and alignment. *Nautilus.*

Medium

1. Hunt, T. (2019). "Resonance and Process Philosophy." *Medium.*
2. Hunt, T. (2021). "Can We Measure Consciousness?" *Medium.*
3. Hunt, T. (2025). "MindFest 2025: My Keynote Talk on Consciousness and AI." *Medium.*
4. Hunt, T. Various articles on consciousness, AI, and political topics. *Medium.*

Additional Resources

Frontiers in Human Neuroscience Research Topics (Lead Guest Editor Tam Hunt)

- "Mapping the Resonome: Electromagnetic Field Patterns in Relation to Cognition and Consciousness" (ongoing). Guest Editors: Tam Hunt, Jonathan Schooler, Christof Koch and others. *Frontiers in Human Neuroscience.*

- "Electromagnetic Field Theories of Consciousness: Opportunities and Obstacles" (completed 2024). Guest Editors: Tam Hunt, Jonathan Schooler, and colleagues. *Frontiers in Human Neuroscience.*

Appendix

Some Further Thoughts on EM Field-Based Healing Devices

As we've explored throughout this book, the therapeutic application of electromagnetic fields represents a fascinating intersection of ancient wisdom and modern technology. This appendix provides practical guidance for readers interested in exploring commercially available electromagnetic resonance devices, from simple frequency generators to sophisticated PEMF systems. While the scientific evidence varies considerably across different technologies and applications, understanding the landscape of available devices can help readers make informed decisions about their own explorations in energy medicine.

The Scientific Foundations

The Schumann resonance, discovered by physicist Winfried Otto Schumann in 1952, represents the fundamental electromagnetic resonance of the Earth-ionosphere cavity, with a primary frequency of approximately 7.83 Hz and several harmonics (14.3, 20.8, 27.3, and 33.8 Hz).[1] This frequency sits at the upper edge of the theta brainwave range (4–8 Hz), a particularly significant placement given theta's role as an "organizing wave" in human cognition and consciousness.[2]

Modern neuroscience has revealed that theta rhythms serve as carrier waves for cross-frequency coupling—a process where higher frequency gamma waves (30–100 Hz) nest within the slower theta oscillations, allowing information to be transported across different brain regions.[3] This theta-gamma coupling is essential for memory formation, spatial navigation, and the binding of distributed neural activity into coherent conscious experience.[4] The correspondence between Earth's fundamental resonance and this crucial organizing frequency in human cognition suggests a deep evolutionary entrainment between human consciousness and our planet's electromagnetic environment.

Recent research supports this connection. Pobachenko *et al.* (2006) found significant correlations between daily variations in Schumann resonance intensity and human EEG patterns, particularly in the theta band.[5] Cherry (2002) proposed that human circadian rhythms, melatonin production, and even higher cognitive functions may be influenced by Schumann resonance signals, which could explain why many people report cognitive disruption in electromagnetically shielded environments.[6] The theta frequency's role as a master coordinator—organizing faster beta and gamma oscillations that carry specific information content—makes the 7.83 Hz Schumann resonance particularly relevant for devices aimed at optimizing consciousness and cognitive function.

Field Strength Considerations

A critical factor often overlooked in consumer electromagnetic devices is field strength. FDA-approved PEMF devices typically generate magnetic fields ranging from 0.1 to 30 milliTesla (1,000 to 300,000 microTesla), sufficient to induce measurable currents in biological tissue according to Faraday's law of electromagnetic induction.[7] In contrast, many consumer "frequency generators" produce fields measured in nanoTesla or picoTesla—millions of times weaker than clinical PEMF devices and often weaker than the electromagnetic noise in a typical urban environment.[8]

However, the role of field strength becomes more nuanced when considering resonance affects on neural oscillations, which might allow even very small signals to exert noticeable affects. Some researchers argue that biological systems may respond to extremely weak fields if they are at biologically significant frequencies, particularly when these match the brain's organizing frequencies like theta.[9] This resonance hypothesis suggests that even weak signals at the right frequency might influence neural synchronization, though this remains inadequately tested.

Categories of EM Healing Devices

What follows is a brief survey of the various kinds of EM healing and therapy devices.

1. Simple Frequency Generators

These devices, typically priced between $40–200, generate continuous or pulsed electromagnetic fields at user-selectable frequencies. Common examples include:

- Schumann resonance generators: Fixed or adjustable around 7.83 Hz to entrain theta rhythms
- Multi-frequency generators: Adjustable from 0.01 Hz to several MHz
- "Sacred frequency" devices: Pre-programmed with Solfeggio frequencies (396, 417, 528, 639, 741, 852 Hz)

The scientific rationale for Schumann generators is strongest, given the documented role of theta as an organizing frequency for consciousness. EEG studies demonstrate that external theta stimulation can enhance memory consolidation and facilitate the theta–gamma coupling necessary for cognitive binding.[10] However, most consumer devices produce fields far weaker than those used in laboratory studies, raising questions about their efficacy beyond potential placebo effects.

2. Rife Machines and Frequency-Specific Devices

Royal Raymond Rife claimed in the 1930s that specific frequencies could destroy pathogens through mechanical resonance, similar to an opera singer shattering a wine glass.[11] Modern "Rife machines" typically operate at frequencies ranging from 20 Hz to several MHz, with databases claiming specific frequencies for various conditions. These devices vary widely in price ($200–5,000) and sophistication.

The scientific evidence for Rife's original pathogen-destruction claims remains absent from peer-reviewed literature. However, some researchers have found that certain frequencies can influence cellular processes. Zimmerman *et al.* (2020) demonstrated that specific electromagnetic frequencies could affect bacterial growth rates, though at field strengths far exceeding those of consumer Rife devices.[12] The FDA has not approved any Rife-type device for medical treatment and has taken action against companies making therapeutic claims.[13]

3. Clinical-Grade PEMF Systems

FDA-cleared PEMF devices represent the most scientifically validated category of electromagnetic therapy devices. These systems, ranging from $500 to $30,000, generate magnetic fields strong enough to induce therapeutic currents in tissue:

- Bone growth stimulators: FDA-approved since 1979, using 1–100 Hz at 0.1–20 mT[14]
- Depression treatment (rTMS): FDA-approved, using 1–20 Hz at 1–2 Tesla[15]
- General wellness devices: FDA-registered Class II devices like BEMER, using complex signals at 3.5–35 μT[16]

The clinical evidence for PEMF is substantial for specific applications. A 2020 systematic review found significant benefits for osteoarthritis pain relief,[17] while a 2019 meta-analysis confirmed efficacy in accelerating bone healing.[18] The key difference between clinical PEMF and consumer frequency generators is field strength —clinical devices generate fields sufficient to induce measurable physiological changes according to established electromagnetic principles.

4. Radionics and "Subtle Energy" Devices

The term "radionics" encompasses devices claiming to detect and transmit "subtle energies" for diagnosis and healing. Originating with Albert Abrams in the early 1900s, radionics has evolved from mechanical devices to digital systems like the Spooky2 and various computerized "quantum biofeedback" machines.[19]

From a scientific perspective, radionics lacks any validated mechanism of action. The FDA classifies radionic devices as fraudulent when marketed for medical purposes.[20] However, these devices maintain a following in alternative health communities, with users reporting subjective benefits that may be attributed to placebo effects, relaxation responses, or the therapeutic value of intentional self-care practices.

Practical Considerations for Device Selection

For readers interested in exploring electromagnetic devices, I offer these evidence-stratified recommendations:

Strongest Evidence:

- FDA-cleared PEMF devices for specific conditions (bone healing, depression, pain)
- Clinical TMS systems (requires medical supervision)
- Price range: $2,000–30,000

Moderate Evidence:

- Consumer PEMF devices with measurable field output (>100 μT)
- Schumann resonance generators for meditation/relaxation (acknowledging theta entrainment potential)
- Price range: $200–2,000

Weak/Anecdotal Evidence:

- Multi-frequency generators
- "Rife machines" (for experimental/exploratory use only)
- Price range: $100–1,000

No Scientific Evidence:

- Radionics devices
- "Quantum" biofeedback systems
- "Subtle energy" transmitters
- Price range: Varies widely

Red Flags to Avoid

Be skeptical of devices or companies that:

- Claim to cure serious diseases (cancer, HIV, etc.)
- Discourage conventional medical treatment
- Use pseudoscientific terminology ("quantum healing," "scalar waves")
- Lack specifications for actual field strength output
- Cost more than $5,000 without FDA clearance

- Promise results for a vast array of unrelated conditions

Starting Your Exploration

For those interested in personal experimentation:

1. Begin with entrainment: A simple 7.83 Hz Schumann generator ($40–100) can provide direct experience with frequency effects on consciousness, particularly during meditation when theta states are naturally enhanced.
2. Document your experience: Keep a journal noting subjective effects, time of day, duration of use, and any other relevant factors.
3. Consider field strength: If you experience benefits from weak-field devices, consider whether stronger clinical-grade PEMF might provide greater effects.
4. Maintain skepticism: Remember that powerful placebo effects can occur, especially with novel interventions involving technology and ritual.

The Consciousness Connection

Perhaps the most intriguing applications of electromagnetic devices relate to consciousness exploration rather than medical treatment. Given that consciousness appears to involve synchronized electromagnetic oscillations—with theta serving as the organizing frequency for binding higher frequency bands—devices that can entrain these rhythms may offer tools for exploring altered states and optimizing cognitive function.

The 7.83 Hz Schumann frequency's position at the upper edge of theta makes it particularly interesting for consciousness applications. During deep meditation, the brain naturally increases theta power, and this theta rhythm serves as the scaffold upon which faster oscillations representing specific thoughts and perceptions are organized.[21] External entrainment at this frequency might facilitate the natural organizational processes of consciousness, though this remains speculative pending further research.

Conclusion

The landscape of electromagnetic healing devices spans from scientifically validated medical instruments to purely speculative "energy medicine" machines. While FDA-approved PEMF devices have demonstrated efficacy for specific conditions, most consumer frequency generators produce fields too weak to induce direct physiological effects through conventional electromagnetic mechanisms.

However, the possibility of resonance effects—particularly with the theta-organizing frequencies of consciousness—suggests that even weak fields at the right frequencies might influence neural synchronization and cognitive states. The correspondence between Earth's Schumann resonance and the brain's theta organizing frequency is particularly intriguing, hinting at deep evolutionary connections between human consciousness and our electromagnetic environment.

As with any healing modality, the integration of intention, ritual, and belief can produce therapeutic benefits independent of the proposed mechanism. For consciousness exploration and optimization, electromagnetic devices offer tools that may work through both validated electromagnetic effects and the powerful influence of directed awareness. The key is maintaining both open-minded curiosity and critical evaluation as we explore these frontiers where ancient wisdom meets modern technology.

References

1. Schumann, W.O. (1952). "Über die strahlungslosen Eigenschwingungen einer leitenden Kugel, die von einer Luftschicht und einer Ionosphärenhülle umgeben ist." *Zeitschrift für Naturforschung A*, 7(3–4), 149–154.
2. Nunez, P.L., & Srinivasan, R. (2006). *Electric Fields of the Brain: The Neurophysics of EEG*. Oxford University Press.
3. Canolty, R.T., & Knight, R.T. (2010). "The functional role of cross-frequency coupling." *Trends in Cognitive Sciences*, 14 (11), 506–515.
4. Lisman, J., & Jensen, O. (2013). "The theta-gamma neural code." *Neuron*, 77(6), 1002–1016.

5. Pobachenko, S.V., *et al.* (2006). "The contingency of parameters of human encephalograms and Schumann resonance electromagnetic fields revealed in monitoring studies." *Biophysics*, 51(3), 480–483.
6. Cherry, N. (2002). "Schumann resonances, a plausible biophysical mechanism for the human health effects of solar/geomagnetic activity." *Natural Hazards*, 26(3), 279–331.
7. Markov, M.S. (2007). "Pulsed electromagnetic field therapy: History, state of the art and future." *The Environmentalist*, 27(4), 465–475.
8. Panagopoulos, D.J., Karabarbounis, A., & Margaritis, L.H. (2002). "Mechanism for action of electromagnetic fields on cells." *Biochemical and Biophysical Research Communications*, 298(1), 95–102.
9. Adey, W.R. (1993). "Biological effects of electromagnetic fields." *Journal of Cellular Biochemistry*, 51(4), 410–416.
10. Reinhart, R.M., & Nguyen, J.A. (2019). "Working memory revived in older adults by synchronizing rhythmic brain circuits." *Nature Neuroscience*, 22(5), 820–827.
11. Rife, R.R., & Crane, J. (1950). "The Universal Microscope." *Journal of the Franklin Institute*, 250(5), 415–436.
12. Zimmerman, J.W., *et al.* (2020). "Targeted treatment of cancer with radiofrequency electromagnetic fields amplitude-modulated at tumor-specific frequencies." *Chinese Journal of Cancer*, 32(11), 573–581.
13. FDA Consumer Updates (2023). "Fraudulent Coronavirus Disease 2019 (COVID-19) Products." FDA.gov
14. Bassett, C.A.L. (1989). "Fundamental and practical aspects of therapeutic uses of pulsed electromagnetic fields (PEMFs)." *Critical Reviews in Biomedical Engineering*, 17(5), 451–529.
15. O'Reardon, J.P., *et al.* (2007). "Efficacy and safety of transcranial magnetic stimulation in the acute treatment of major depression." *Biological Psychiatry*, 62(11), 1208–1216.

16. Bohn, W. (2013). "The technological development history and current significance of the 'Physical BEMER® Vascular Therapy'." *Journal of Complementary and Integrative Medicine*, 10(Suppl), S1–S13.
17. Yang, X., *et al.* (2020). "Effects of pulsed electromagnetic field therapy on pain, stiffness and physical function in patients with knee osteoarthritis: A systematic review and meta-analysis of randomized controlled trials." *Journal of Rehabilitation Medicine*, 52(6), jrm00077.
18. Peng, L., *et al.* (2019). "Pulsed electromagnetic fields for fracture healing: A meta-analysis of randomized controlled trials." *Bioelectromagnetics*, 40(8), 547–559.
19. Jensen, D.C. (2019). "A Short History of Radionics." *Subtle Energies & Energy Medicine*, 29(1), 41–68.
20. FDA Compliance Policy Guide Sec. 400.210 (1987). "Radionics Devices." FDA.gov
21. Buzsáki, G. (2006). *Rhythms of the Brain*. Oxford University Press.

Index